Beekeeping
The Gentle Craft

JOHN F. ADAMS

WYNWOOD™ Press
New York, New York

Library of Congress Cataloging-in-Publication Data
Adams, John Festus.
Beekeeping: the gentle craft / John F. Adams.
p. cm.
Reprint. Originally published: 1st ed. Garden City, N.Y.: Doubleday, 1972.
Bibliography: p.
ISBN 0-8007-7200-8
1. Bee culture. I. Title.
SF523.A27 1988
638'.1—dc19 88k-10684
CIP

Published by WYNWOOD™ Press
New York, New York
Printed in the United States of America

For Emily, Erika and Stephanie:
of the 20,000 females
who inhabit our acre,
my unrivaled favorites.

Contents

Beekeeping
The Gentle Craft

1

The Hot-Stove League

POET'S VOICES TO THE CONTRARY, spring and summer are not the seasons of dreams. Between the vernal and autumnal equinoxes, a normally healthy human animal expends his creative energies in creating, not dreaming of creating. Depending largely on his age, he may expend his energies chasing balls, belles, or rose thrips, but according to his capacities he chases. It is easy to confuse the exhilaration of the chase with the reveries of fancy, but the fact remains that for the most part the mellow weeks of spring and summer are spent in doing things rather than thinking about things.

It is that ebbing season between the autumnal and the vernal equinox, through that almost pulseless slackness of the winter solstice, that dreams and dreaming tease us into believing that the sun will indeed shine again, and the animal form again throb with strong pulse and toned muscle. This is the season when despair may become so deep that only a total resurrection of the universe can be believed possible to melt the snow; but it is also the season for the miracle of faith—faith strong enough to believe even in the prophecy of a seed catalogue, which arrives an exact month before Lent.

Even though the worst of the winter is most likely still to come, the dreams awaken. We reminisce about the season past, and gradually discover our reminiscences to focus less on what has passed than what is to come. We find we actually believe in spring. And spring is still far enough in the future that our dreams are not contaminated seriously by considerations, certainly not considerations of practicality. We can actually dream of planting an oak tree where gray squirrels may perch, to realize fantasies one time set off by pictures in a childhood story book. We may dream the dream of the Lake Isle of Innisfree, with its willow-wattle cottage; of the nine bean rows, and the hive for the honeybee.

The true dreamers of us may, that is. In the season of dreams, the lake isle is credible, the wattled cottage almost credible; the bean rows—nine precisely—require a specially willing imagination. But the hive for the honeybee requires a leap of the imagination; not a dream, but a spiritual illumination.

The pre-vernal urge to raise bees does not strike everyone, and those it does strike are smote in wildly differing degrees of seriousness. There is perhaps reason to suspect that there may even be a certain kind of genetic disarrangement in the chromosomes which must account for some of the apiphilia. It is rather to provide some means of self-analysis, to find if the vocation to keep bees is a genuine one, and not just a cruel hoax of the January seed-catalogue syndrome.

It is not quite the same thing to send off in January for the materials of beekeeping as to mail three dollars for a bare root oak tree.

If the apicultural fit falls on you, by all means make fullest use of the winter months for study. Keeping bees is relatively simple, like raising children is relatively simple. Relatively. Dr. Spock does not tell you all you need to know (he may even lie benevolently), but he is more of a comfort and convenience than austere clinical pediatric consultants. Unlike children, hives of bees do not occur by accident. To a considerable degree you can be apprised in advance of the likely difficulties of raising bees, and decide in advance whether the strength of your apicultural impulses are equal to the problems.

Aside from consulting a good book (this one, for example), you should first of all subscribe to at least one bee journal. There are two good American ones, and since subscriptions are not terribly expensive, you would probably do well to take them both. They are invaluable sources of ongoing information about the craft or art of beekeeping, and are in effect complete indexes of advertisements and catalogues. The *American Bee Journal* (Hamilton, Illinois, $11.95 per year), is in its 110th year of continuous publication, and while it is clearly directed to the professional, there is no condescension toward the hobbiest. It is nicely illustrated, and invariably informative. The second important American journal is *Gleanings in Bee Culture* (I. A. Root Company, Medina, Ohio, $12.49 per year), in its 97th year of continuous publication. It is also directed to the professional reader, but its quality is not as uniformly high as the *Bee Journal*. If you can only subscribe to one, the *Bee Journal* is the best choice. There are also numerous English language journals published abroad (and in Canada) which you might eventually want to look at.

In the first issue you examine, you will find advertisements for bees and bee equipment in quantity and variety which will perhaps astonish you. These advertisements will themselves probably answer volumes of real and inchoate ideas of the incipient beekeeper. For example, the first questions a neophyte ever asks are where do you get bees, and how much do they cost. By the time you have read a couple of issues, you will find your interests either waning or subsiding entirely, or else whetted to a feverish pitch of desire. If you find yourself in this latter condition, you should answer a number of advertisements which offer free catalogues for bee supplies. These catalogues, when they arrive, will answer numerous additional questions about complexity, cost of equipment, and finally, sheer practicality, given your circumstances.

During this same period of time you will probably also be examining varieties of books on the subject of beekeeping. Probably no insect has been as extensively studied as the honeybee; certainly none has been written about so divergently. For one thing, you will find the books sometimes at odds, absolutely contradictory at times. And information may vary from the oversimple to the overcomplex. The amateur will often be very perplexed in attempting to tell which is which. This present book is not conceived as offering anything new to the science of apiculture. It is a basic book for the beginning beekeeper, answering questions about *whether* you want to start as well as how to go about starting. Anyone who gets a start in keeping bees should read as widely in the literature of beekeeping as he can, although reading too widely too soon may serve to intimidate the beginner with his first hives. But no matter how absorbed one gets in the subject, it is quite certain he will never master everything which is available to be known about bees and beekeeping; certainly he will never learn enough about them to remove all uncertainty from the practice of beekeeping. The essentials of beekeeping, however, are quite simple; it is the purpose of this book to guide the novitiate through the establishment and maintenance of his first colonies.

You might profitably begin your preliminary study of bees by giving yourself a little psychological test of your suitability—not aptitude—for handling bees. If, in a test of simple word association, your impulse is to pair *bee* with *honey*, you probably may consider yourself suitable; if your impulse is to pair *bee* with *sting*, your suitability is in question. You will at least need some re-education in the latter instance. Bees do sting, and people who handle bees do get stung. There is no technique or esoteric system of beekeeping to prevent bee stings, and no arcane unguent known only to the initiated which will keep stings from hurting. Beekeepers do develop a certain resistance to bee stings through being stung repeatedly, to the extent that a sting seldom swells or even hurts longer than a minute or two; but when the sting goes in, it still hurts about the same regardless of your professional credentials.

Bees do not, of course, "bite." Their mandibles are designed primarily to manipulate wax and carry debris, and are incapable even of tearing the soft corolla of a flower. Bees sting by thrusting a specially designed posterior appendage into your skin. This stinger of the bee is in itself a marvelous and ingenious instrument. An engineer would despair to improve on its design. Essentially it consists of two probing parts, both barbed, and both capable of being activated by the muscles of the base of the abdomen to which they are attached. When the bee homes in, the sting is thrust without hesitation into the skin; the barbs hold the sting so firmly in the skin itself that the base of the abdomen is actually pulled off the bee's body, and together with the poison sac and activating muscle remains against the skin. The bee always dies, and of course can only sting once. The detached muscle continues to contract, working the barbed stinger deeper into the skin, and helping to pump poison into the blood stream.

The moral of this brief lesson in bee anatomy should be obvious. Remove the sting as quickly as possible to keep it from working deeper and to keep it from pumping the full contents of the poison sac into your system. And you should see that you do not pull it out; squeezing the mass at the base of the sting will simply have a hypodermic effect and ensure that you will get the full dose of poison in the process of extraction. Rather, *brush* the sting away. In spite of its barbs, it actually comes out quite easily. Numerous folk remedies and commercial preparations are suggested for application on the circular white weal that springs up almost immediately where the sting went in. (You will find some advertised in the bee magazines and bee catalogues.) Usually I do nothing, unless stung in some part of the face that is inclined to swell. On such stings I usually apply a dab of honey. My grandmother told me this would work when I was a little boy. I really don't know if it works or not, because I have been stung so many times during the course of my life that I don't swell much anyway. You may want to try this medication; it works at least as well as any commercial preparation I've tried, since they don't seem to work at all.

Bee venom is real venom, akin in its constitution and operation to snake venom. Research on bee venom was retarded for some years because of an offhand comment many years ago by an expert on bees to the effect that the toxic agent in bees' stings was formic acid, the same as ants'. This mistake perpetuated itself a considerable time, with research being conducted with formic acid from more readily available sources than the venom in bee stings.

Some people have a serious allergy to bee stings. Such people obviously should avoid bees. You absolutely may not keep bees without occasionally getting stung.

It is true that experienced beekeepers are less frequently stung than beginners. This does not result from any camaraderie that may be established between keeper and kept, or from "taming" the bees. The experienced bee keeper will learn to judge the temper of a hive before opening it, as well as to continue to gauge his bees' tempers while he has the hive open. He will also master the use of the smoker, the convenient hand-held device for administering calming smoke, when and where to use it, how much, and when is *too* much. But above all, the experienced bee keeper will learn to handle the hives, frames, and bees more smoothly and gently than the beginner, so the bees won't bother him because he doesn't rile them up. And a final factor; bees in a well-managed apiary are by breeding kept as gentle as possible.

Protective clothing of course is worn by most beekeepers, at least occasionally. Bees seem to be calmer if white is worn, or at least light-colored clothing, and are less disturbed by cotton or linen than animal or synthetic fibers. This is perhaps because they are used to brushing plant fibers, are put on guard or aroused by animal substance. Protective gloves are sometimes worn. Standard bee gloves are clumsy and make handling frames awkward, and of course tend to make one handle all the equipment roughly, and particularly to pinch and squash bees. Bees sting the gloves, and the smell of the venom serves to invite the rest of the bees, regardless of their previous disposition. Eventually gloves which are used regularly will become permeated with the smell of bee venom, so that the gloves alone may be sufficient to anger or excite the colony. The mark of the real professional beekeeper is wearing gloves to protect one's hands from stains, not stings.

According to a number of prevalent and traditional notions, bees have certain peculiar but universal peccadilloes. One such notion has it that bees are aroused if you try to work them after having an alcoholic drink. Those people I have heard express this idea with most conviction have been people most unlikely to ever have had personal experience with alcohol. I am not such a person, but nevertheless have never tested the idea because I never work bees after the sun drops below the yard arm, and before that time never tipple. Similarly, I've heard it said that bees resented the odor of certain foods, garlic and onions, for example. There is a prevalent superstition that bees are offended by the presence of women during their menstrual period, an idea which suggests an origin in Deuteronomy or Leviticus. There is even a superstition that bees don't like people with red hair. This last notion, at least, I can speak to with authority. I have been of the red-haired persuasion all my life, and can't say I've ever noticed bees to be more uncongenially disposed toward me than to anyone else. It would perhaps dispel centuries of superstition and misunderstanding about bees if I would go into the apiary dead drunk, reeking of garlic and onions.

Speaking of the area of superstition—and there are an extraordinary number associated with bees and their keeping—many may be attracted to the hot-stove league by notions that to keep bees is a specific remedy or preventative of such ailments as rheumatism, arthritis, hay fever, etc. Some say it is the sting of the bee which is efficacious, some believe the effluvia of the hive are healthful. Some attribute the curative powers (especially for hay fever or asthma) to the honey itself. Proponents of these claims point to the strong physiques and rugged constitutions of practicing beekeepers. Maybe so—but of course unless you are reasonably sturdy to begin with, you'll find the physical work of handling bees a considerable burden. There are few decrepit beekeepers because decrepit people can't lift beehives. Nevertheless, some of these claims or potential properties of bees and honey are the objects of extensive and ongoing research.

Additional claims are also made about the healthy properties of honey. This, on general principles, no sane person would dispute; anything that good has to be either absolutely healthy or totally immoral. If you want to indulge in fantasies about its properties, even that honey is a super-rejuvenant, there is no harm in doing so. These are, if fantasies, ones in which I would like to believe.

There are also claims made about "royal jelly," the name given to the specialty food secreted to the queen bee which allows her to develop into such a remarkably different bee from her sterile sisters. There have been in the past what almost amounted to fads associated with the marvelous properties of this substance when eaten by humans. As far as I know no real proof of its effects one way or another on humans has been established. You will, however, see it advertised in the bee journals. Its cost should astonish you.

If you passed the "word association test," believe in The Bee, and have subscribed to the bee journals, you are in imminent danger of becoming an apiophile before the hot-stove league has advanced beyond the stage of adding columns of figures on scraps of paper. By the end of February your enthusiasm may have soared you into dreaming mastery over hundreds of hives, numbering millions of individual subjects, working day and night in your interest and to your profit, not to mention your pleasure. It is possible to make a living from bees, but don't be deceived by the appearance that somehow the flowers and nature combine to give freely your living. Nature and flowers aren't built that way, in case you haven't noticed.

First off, notice the price of honey in your supermarket. It's cheap by any standards, almost criminally cheap; when you think of the bee-miles represented by a five-pound can of honey, it should shame you to pay such a pittance. Then look at the monthly honey reports in the bee journals—production and wholesale prices. Honey prices have varied remarkably little in recent history; currently it wholesales at about forty cents a pound, depending on quality. To make a profit at such a price requires huge production, which requires extensive capital outlay. Honey production does provide the livelihood of many people. But an amateur shouldn't think of trying to make money, even modestly, from his bees, at least not for several years. There is a story of a man asked what he would do if he inherited a million dollars. He thought for a while and responded, "I would keep bees for as long as the money lasted."

2

Starting a Colony

Without weighing such serious implications as the facts of life about bees or the metaphysics of bees (topics for later chapters) during some given winter the madness of fancy finally drives you to the firm decision to try to take up keeping bees. The very time you decide is probably not too soon to take definite steps toward getting your supplies on order. If you have subscribed to the journals and ordered some catalogues from bee supply houses, the mystery of where such esoteric furniture as bee hives and bees themselves originate has been dispelled. You may also, incidentally, order most of the necessities (including the bees) from such domestic mail order houses as Sears Roebuck. Simply request their farm catalogue for information. Most things you order from Sears will come prepaid, which is frequently worth considering, especially when ordering hives and supers, the wooden "stories" added above the main hive, in which the bees store recoverable honey.

Your first decision, after the initial one of affirmation, is how many colonies to start with. The timid will think in terms of one; the expansive type—the potential entrepreneur—will access his ambition in terms of six or eight for starters. It is well to begin small, but not too small. For reasons to be advanced later, you should have at least two colonies initially, but as a beginner probably no more than three. Cost may well be the factor which makes the decision of numbers; a single hive body will cost, complete and ready to receive bees, about fifty dollars; a three-pound swarm of package bees about twenty-five dollars. Additional supers needed to be added to the hive as the season progresses will cost an additional minimum of around fifteen dollars (one deep super and two shallow supers, together with their frames and wax foundation). In other words, you must plan on laying out about ninety dollars on your first colony.

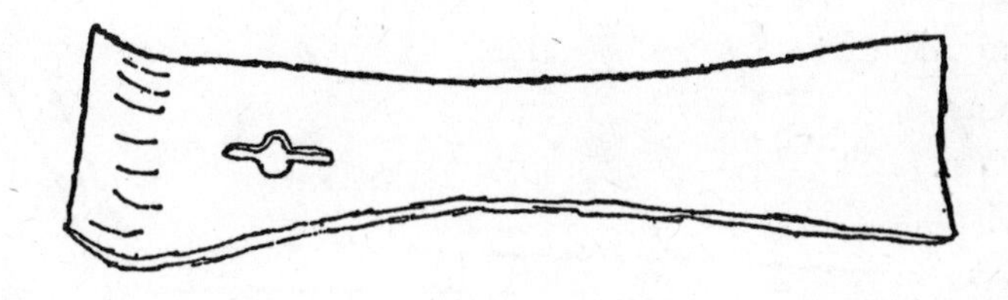

Hive tool

By way of additional equipment you will also need as an absolute minimum a bee veil, a pair of bee gloves, a hive tool, and a smoker. Some supply houses package what they call a basic beginners outfit, including these items (and usually a bee book thrown in) for around forty dollars. This set may be a good bet for a beginner; remember,

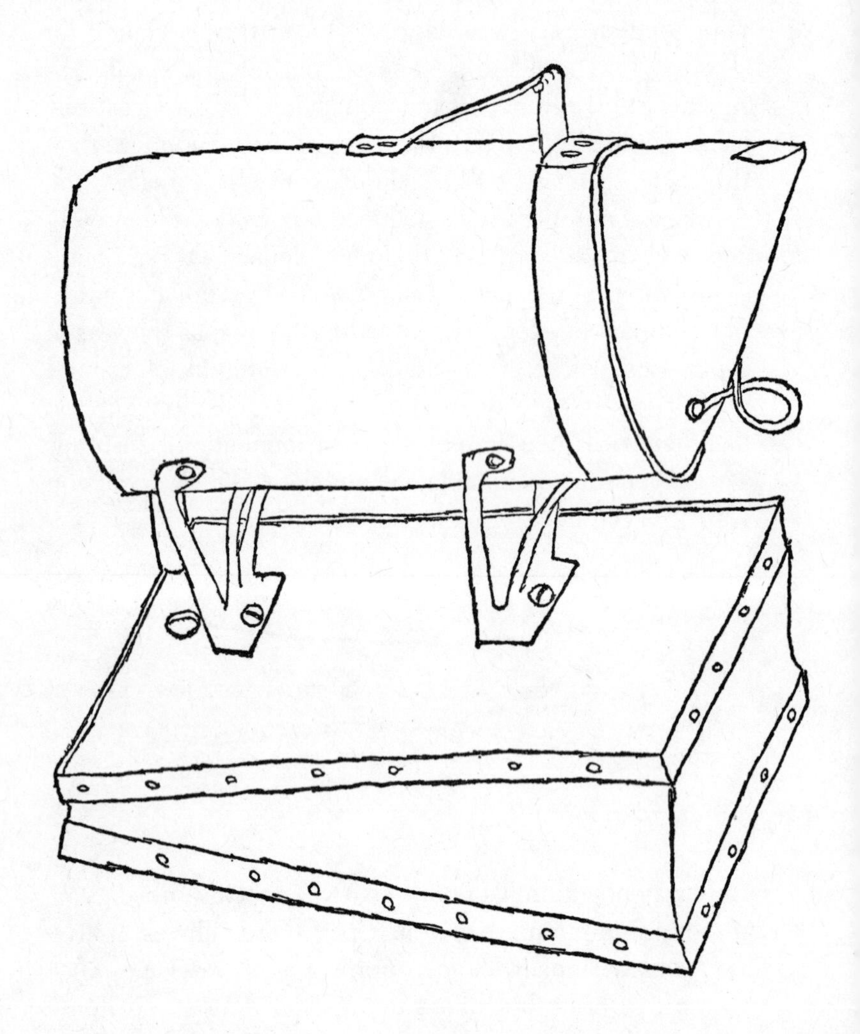

Bee smoker

however, that with this basic outfit you are still going to have the expense of "supering" before the summer ends, which will cost an additional twenty dollars or so. Two complete colonies, then, will cost about $160 for the first year. The expenses for your second year, providing you do not expand, will of course be much reduced; no more than an additional twenty dollars to maintain the two colonies, perhaps as little as ten.

But I hear a question being formulated in the back of the room—"doesn't the number of colonies you need to start with depend on how much honey you intend to produce?" Oh yes; ultimately we all envisage a honey harvest. However, if you are planning to amortize your expenses by selling honey, or at least justify to yourself or your wife the expenses of establishing your apiary by the *prospect* of selling honey, you are adding an unstable variable to the whole science which can kill the project before it germinates. At the very least pecuniary considerations, advanced too soon, can appreciably dilute the joy of prospective beekeeping. Practically speaking, if your bees arrive on time, and everything proceeds well in their establishment, and you don't pester them too much, if there is a good honey flow in your area you ought to harvest eighty pounds per hive the first year. It may well be a good deal less. Some northern beekeepers, Canadians particularly, never winter their swarms, but start out fresh each spring with package bees shipped in from the south and still harvest 120 pounds and more per colony. They have excellent forage for their bees, however, and install their package bees in hives filled with drawn comb. And they are very skillful beekeepers. All I can really advise an amateur on this point is that two established colonies are a good provider for a single family, and it is unlikely your immediate family will consume what your two swarms will normally produce. Don't, however, plan on supporting your expenses by selling a surplus. I have managed up to twenty-five colonies, and never sold the contents of a single golden cell. But then, I like honey and am friendly with a lot of people.

When you begin with your first colonies of bees you cannot afford to be frugal, and by no means resort to makeshifts or cobbled-up hives and equipment. It may be that you have access to some used hives and frames, or know of some old colonies in which the bees have died out. Avoid these windfalls like the plague, because that is quite literally what they will give you—*plague,* in the form of one or another deadly bee disease, the dreary topic of a later chapter. You may be able to catch a "wild" swarm of bees to inhabit your waiting hive, but that is a matter of advanced serendipity, and more likely will occur later in your beekeeping incarnation than earlier.

At this initial point, let's assume you will begin from scratch. You will need, if you project a two-colony beginning, two standard hives, complete with reversible bottom boards and tops, preferably metal covered. You may order your supering equipment at the same time you order your hive, or you may order it later; in any case you will not apply the supers to the colonies immediately.

The bee journals and the catalogues will tell you most everything you need to know about the styles and sizes of supers and the frames which fill them to contain the combs. In America, frame sizes are completely standardized, and the major dimensions of the hives themselves are mostly standardized. The standard American beehive holds ten frames (although some beekeepers use only nine frames). There is also a design which holds eight frames, largely obsolete; it may be had through many suppliers simply by specifying the size (the price is the same as for a ten-frame size). Many of the older apiaries are equipped with this size, and it will often be seen in the countryside. Whatever size you choose, stick to it so that all your equipment will be interchangeable. While some beekeepers are militantly of contrary opinion, I recommend using the standard ten-frame hive.

There are three sizes of super, $9\frac{9}{16}$ inches deep (call it 10 inches), $6\frac{5}{8}$ inches, and $5\frac{11}{16}$ inches. The 10-inch super is the depth of the standard hive body, and many commercial beekeepers use this size exclusively, for supers as well as hive bodies. Others use a 10-inch hive body, with a $6\frac{5}{8}$-inch super as a second-story brood chamber. Some use both the $6\frac{5}{8}$- and $5\frac{11}{16}$-inch supers for extracting honey. There are certain advantages to the middle-sized super, but it is disproportionately more expensive because the odd size does not match any standard dimension of lumber. If you are producing comb or chunk honey, you should use the $5\frac{11}{16}$-inch super; for one thing, this size enables you to produce the whitest, clearest comb because the bees fill it faster and it is therefore not subject to the normal discoloration of extended traffic by the feet of bees, mundane as that may seem. For the beginner I would advise using two 10-inch supers, one for the basic hive body, and the other for a second brood chamber; to super

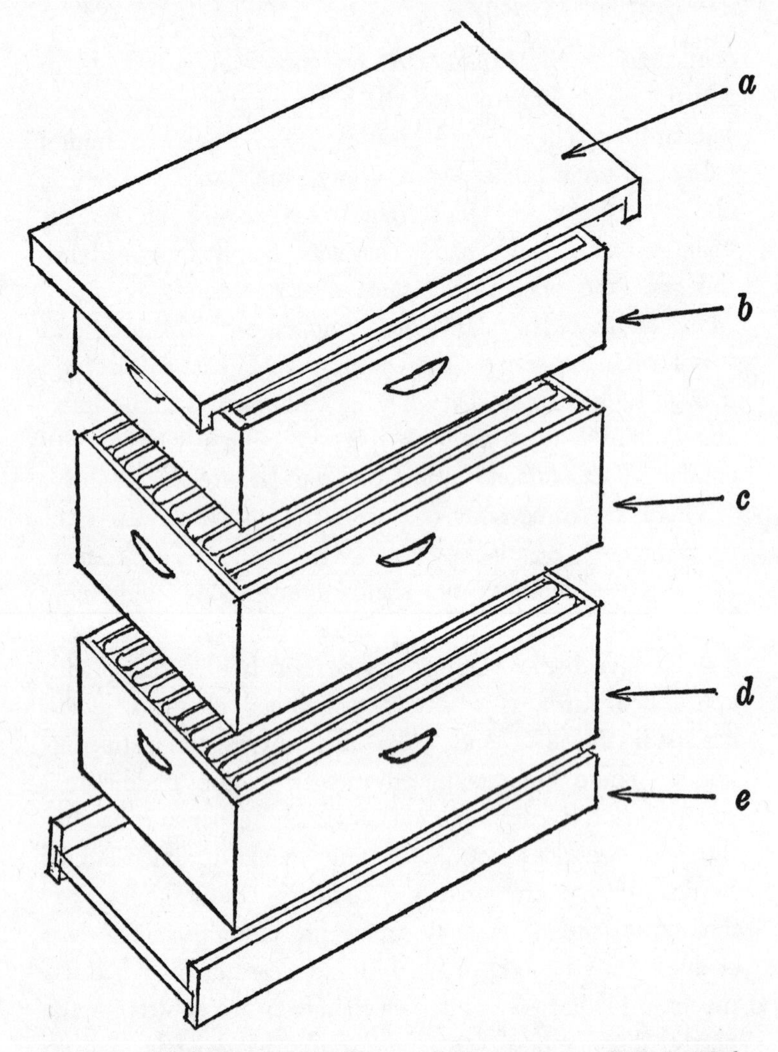

A two-brood chamber hive with one shallow super a. *Lid* b. *Shallow super* c. & d. *Deep super brood chambers* e. *Reversible bottom board. To narrow hive entrance, turn bottom board upside down and replace in hive.*

for honey I would recommend using the 5¹¹⁄₁₆-inch depths. Whether or not you decide ultimately to begin producing extracted honey, this depth will be suitable, and very convenient. But by that time you will have become an experienced beekeeper and developed your own unbearable prejudices on the subject.

The hives which you order will arrive "knocked down" (which explains the initials "k.d." on the catalogue, if you were curious). The materials of the hives are usually pine or cyprus, and the side pieces are dovetailed, with a small hole drilled at each point a nail is to be driven, which prevents splitting. Follow the directions for assembly that accompanies the hive; they are usually almost unintelligible, but you can go badly wrong if you don't make an effort to understand them. By all means fit together all of the dovetailed edges of the hive (making certain these slots face each other) before you begin to nail, making certain that the hand holds on all four sides face *out* and *up*, and that the recessed edges, front and back, are facing each other on top. These recesses form the shelves from which the frames are suspended. Get the slots to fit together as tightly as possible, and then begin to nail. I always nail from the center to the sides, placing only one nail at a time, rotating the hive body as each nail is placed. This procedure helps keep joints tight and prevents the awkward discovery halfway through that one side is leaking daylight.

When the hive, top and bottom, is assembled, paint it. For the hobbiest, any good quality white outdoor paint will do; for the sake of convenience, I use outdoor latex. Commercial beekeepers often cut corners on the cost in

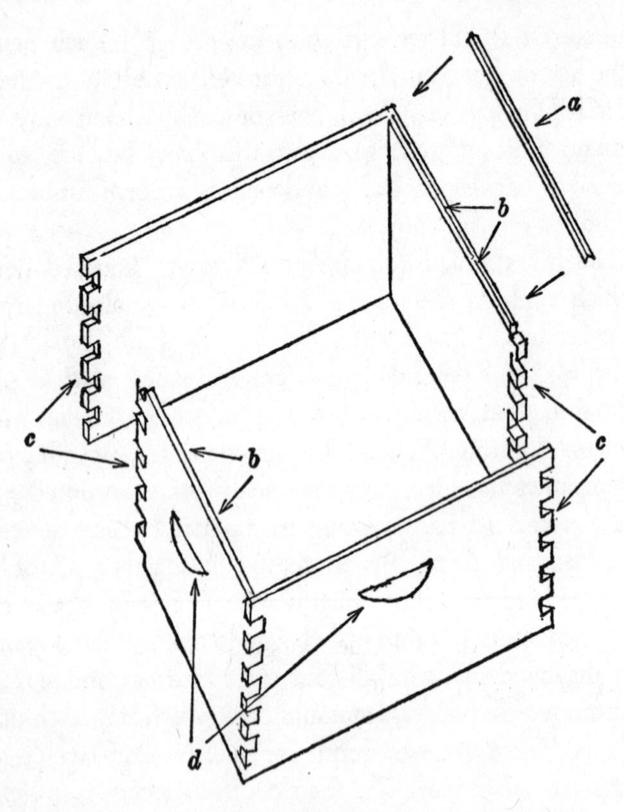

Assembly of the hive body a. *Metal rabbet, nailed into position* b. *Recessed slot upon which frames hang. Metal rabbet helps keep frames from being cemented to the wood.* c. *Slots* d. *Hand holds*

hive painting, but certainly the hobbiest shouldn't. Give the hive three coats. Do not paint the *inside* of the hive, the *underside* of the top, or *either* surface of the bottom board. To do so would outrage your future bees.

You must, of course, arrange the orders for your material so that your hive arrives some time—weeks or months—

before your bees. This will give ample time for any paint smell to have dissipated before you install the bees. Never put bees into a freshly painted hive; they might just up and walk away. This is known as *absconding,* but there is small comfort in knowing the word after the fact.

When you have your hives constructed and painted you should assemble your frames. They are very simple to put together, but study them before you begin nailing. You should, by the way, specify when you order the hives that you want them furnished with split-bottomed frames, and when you order additional frames, order the same split-bottomed model. There will be arguments from professional beekeepers on this, but for the beginner, I should think there is no question. Most people actually use this kind of frame now, and except rarely this is the model now most usually supplied with a hive or super. Like the hive, the frames also are usually made of pine. The material is quite thin, even fragile, and must be nailed together with care to prevent splitting. Usually spare side pieces are supplied with an order of frames to make up for a certain percentage of breakage in the assembly process. Advanced beekeeping nuts contrive incredible modalities for assembling their frames; these will occur to you soon enough without suggestions being served in advance.

When you have finished putting the frames together (as you finish each frame, the most convenient and secure way of storing it is to hang it in one of the completed hive bodies) you are ready to install the "wax foundation" in them. Foundation consists of thin sheets of beeswax which is imprinted on each side with the base of the hexagonal honey cell. The bees eagerly seize on this as a "foundation" for erecting the complete cells of the comb. Using foundation provides an enormous labor-saving service to the bees, and comb built on the foundation is much stronger than comb which is drawn free; that is, comb which is made without any guide or support. The whole convenience of the movable frame hive depends on having the combs constructed inside the frames, and there is no other sure way to make certain the bees will co-operate in this convenience than providing them with frames established with foundation. If the frames were to be installed empty, the bees would construct comb any which way, which might be good for their own psyches, but is devastating for hive management.

Foundation comes in a variety of forms. You will, of course, order foundation to fit the depth of your frames (there are three standard frame sizes, corresponding to the three standard super sizes, plus the frames and foundation which produce the small boxes of comb honey sold commercially). Regardless of size, a number of different options are available in the form and size of the foundation which you buy. First of all, if you are using frames with split bottoms, as recommended above, order foundation for split bottoms. The frames you are now assembling in the first stages of your tooling for beekeeping will become the brood chambers of your hive. Consequently, you want to make every effort to give the frames as much permanence as possible. The choices of foundation for brood comb are: wire-reinforced, plastic-reinforced, and unreinforced heavy wax. Wire reinforcement is provided by a series of parallel vertical wires, embedded in the wax about two inches apart or slightly more. Sometimes foundation is extra-reinforced by using a grid of horizontal and vertical wires. Some beekeepers (most commercial beekeepers) use plain foundation and embed their own wire into the wax, using the small holes you will notice in the sides of the frames to support the wires. Some use reinforced foundation *plus* their own horizontal wires. Simplest to use (by a small margin over the wire-reinforced) is that reinforced with plastic. This foundation is composed of a very thin sheet of clear plastic with a thin coat of beeswax on either side, imprinted in the regular manner with the hexagonal base of the honey cells. There are arguments against using plastic reinforced foundation, but personally I prefer it for the brood chambers. It is slightly more expensive than wire-reinforced foundation, which is in turn appreciably more expensive than plain foundation.

There are additional grades of foundation, the differences stated chiefly in terms of thinness and the quality of wax. The very thinnest, made of superior grades of wax, is preferred for producing honey which is to be eaten in the comb rather than extracted. Coarser, that is *thicker*, grades of foundation produce a rib in the center of the comb of finished honey which is tough, and sometimes of a less attractive color than the rest of the comb. If you intend to produce comb honey, it pays to buy the superior grades of thin foundation. If you are extracting your crop and reusing the comb, you will want to use a heavier foundation because it helps make your combs more permanent.

Inserting the comb into the frames is not difficult if you use frames with split bottoms and "wedge" tops. Simply remove the wedge (which is sawn nearly through) from the top of the frame with a sharp knife and start three nails

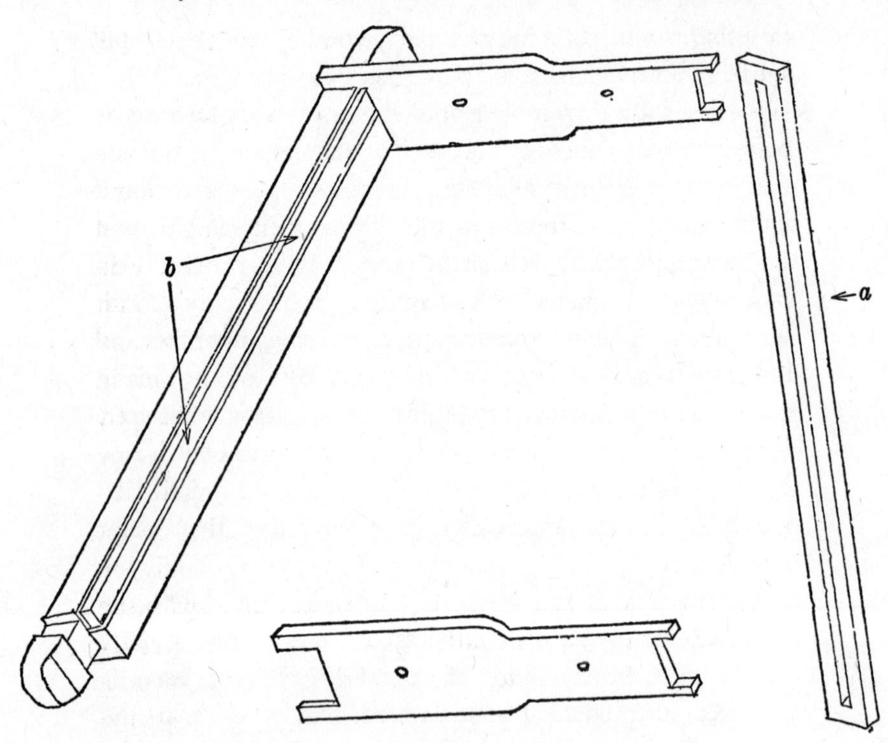

Assembling a frame a. *Split bottom* b. *Wedge*

into the wedge itself. Slide the foundation through the split bottom until it rests along the top of the frame in the recess from which the wedge was removed. Place the wedge back in position and drive in the three nails you have started. Then, holding the frame upright, smoothing the foundation until it hangs perfectly straight, with no bulge or sway, pinch the split bottom together and hold it secure while you drive a single nail across the center of the split bottom. When you have finished putting ten frames together with their foundation for each hive, you are ready to install the bees.

The bees should have been ordered in advance (at least by February) specifying your preferred date for delivery. If you place your order early enough, the bees should arrive about when you ask for them, weather permitting. (Cold, wet weather in the South will delay packaging.) They will ordinarily come by parcel post, as unlikely as that may sound, packed in wire boxes or cages having screen on two sides and a perforated can of syrup secured inside to feed them in transit. Usually you will order a specific quantity of bees, by weight—two to six or seven pounds. I recommend ordering six pounds, although commercial beekeepers would disagree with me for commercial reasons. Many shippers supply only a single standard weight, three pounds. Ideally the arrival of the bees should be timed to correspond with the first blossoms in your area, although this exact time is frankly difficult for a beginner to gauge. Until you begin keeping bees you will never look at flowers with the proper kind and degree of appetite. If you can remember when the dandelions begin to bloom in your particular area, that is a reasonable guide. Dandelion blossoms are in most northern areas an excellent pollen source in early spring, and are very fine to establish a colony on. You should have your bees installed at least by the time of the fruit blossoms.

TYPES OF BEES

Before ordering you should also decide on the variety of bees you want to keep. There are no honeybees native to North America, and all "domesticated" bees are either Old World varieties or hybrids—i.e., first crosses—of Old World varieties. Perhaps you have heard of South American "stingless bees"; they have never proven satisfactory for apiculture. The little heathens are lazy. You have perhaps also heard of African "killer bees." They are now a serious problem in South America, and their importation to this country is strictly forbidden. By far the most prevalent domestic bee in this country is that variety called the Italian bee. It is the common yellow banded one always pictured, and commonly seen, in industrious poses on blossoms. This bee has the advantage of industry, adaptability to the movable frame hive, reasonably good temper (depending on the strain), and resistance to disease. Its genealogical origins are unknown, but it is assumed to have resulted from natural crosses between Egyptian and/or Cyprian bees, probably with Carnolian. This Italian strain is genetically dominant; that is, a queen from another species which mates with an Italian drone will have offspring with Italian characteristics.

Another strain sometimes kept is the Caucasian, originating in the Caucasus Mountains. It is about the same size as the Italian, but grayish in color rather than yellow. The principle advantages of this strain are docility and ease in wintering (this latter probably an advantage developed through the rigors of its place of origin). It is in some ways apparently less adaptable than other strains to the movable frame hive, and sometimes has bad swarming tendencies. At certain times of the year it has the additional bad habit of daubing everything in the hive with gobs of propolis, a kind of bee glue which is frequently a source of annoyance to the beekeepers. There is some evidence that the first bees introduced to the west coast of the United States were the Caucasian strain, being imported from the Russian missions in Alaska to the Russian missions on the California coast. Presumably all traces of this initial importation have long since disappeared. This is a strain which you can buy, or produce by requeening your existing strain, through breeders who advertise in the bee journals.

The Carnolian strain, originating in Switzerland, or Hungary, is an ancestor in many European strains. Its habits of swarming to the point of debilitating the parent hive renders this an undesirable strain (although very desirable in the days of skep beekeeping, when the increase through accumulating small swarms was the essence of beekeeping). It, also, is a grayish bee.

There are various strains of European "black bee," which are seldom or never to be found in this country. The English black bee apparently suffered extinction during an outbreak of an ailment called "acarine infestation" or "Isle of Wight disease" this century. Many British beekeepers now breed a variety of Dutch black bee which is very similar to the native bee. Two other strains, the German black bee and the French black bee are infamous for their bad tempers. The Cyprian bee and the Egyptian bee, mentioned earlier as possible ancestors of the Italian, are also noted for their evil dispositions. It should content the amateur beekeeper simply to know that these strains just discussed exist, and to console him if he has happened to have an unfortunate day with his own hives. Things could be worse. There is also a distinct strain of bees in India, which is little known outside its native ranges.

Within the realm of practicality, there are also hybrid bees. Essentially these are all first crosses of the Italian strain and another, often Caucasian. The hybrids are bred to emphasize the best qualities of different strains, and there is much to recommend them to the amateur. But as is often true of other animals, while first crosses may emphasize the best qualities, second crosses often tend to reproduce less desirable traits. If you try to keep a queen which has been reared in your own hybrid hives you may discover bees more ill-tempered than either of the parent strain, less productive of brood or honey, or with other disqualifying features. If you start with hybrids, most likely as an amateur you should maintain your strain by purchasing new queens from the breeders when requeening is necessary. I have had experience only with the so-called "midnight hybrid," an Italian Carnolian cross, and recommend it with some enthusiasm. They are very gentle (recommendation enough to most beginners), quite productive, and seem to have superior wintering qualities. They may be slightly more prone to swarming than Italians, but seem generally to adapt well to hive management. They are, by the way, industrious producers of propolis. Spring cleaning may be a messy job. If you buy midnights to begin with, they will cost somewhat more than Italian, but I think they are worth it. They are easily distinguished from the predominately yellow-and-black-banded Italians by their grayish bodies. If for no other reason, sweetness of disposition would recommend them to the beginning beekeeper.

3

The Essential Honeybee

THE COMMON HONEYBEE, the *apis mellifera,* has been perhaps the most intensively studied of all insects. Her usefulness of course invites that she be studied, but its unique social organization would probably attract intensive study even without other considerations. Aristotle had an observation hive (presumably set up with isinglass sides to view the activity on the interior), and he is responsible for some of the earliest systematic observations of their social life. Virgil also celebrates the honeybees he kept, notably in the *Fourth Georgic,* and evidence of systematic beekeeping is abundantly represented by ancient art as well as literature. An alcoholic drink, second only in importance to wine and beer (more important than either in some regions) was a beverage made of fermented honey called *mead.* And of course from the Greek root meaning "honey" we have not only the words *mead,* but *mellifluous, melody,* and the name of the insect itself.

The universal western symbol of honey and the honeybee is the conical-shaped straw hive called a *skep.* The expression "beehive shaped" even today denotes that configuration, and some people are still surprised to learn that bees are no longer maintained in those establishments.

For all their picturesqueness, straw skeps were actually a rather barbarous method of beekeeping, since for all practical purposes it was necessary to destroy the swarm in order to harvest its honey. No practices of bee or hive management as we know them today were possible, although some forms of two-story skeps were in occasional use. Among skepists, as those who maintained bees in skeps were called, bees with a proclivity toward swarming were especially valued (and this perhaps accounts for the excessive swarming traits today of some races of bees under long domestication). Only a few colonies were wintered, to serve as a starter for the apiary next year. It was also the days of skep beekeeping that gave us the rhyme:

> A swarm of bees in May
> Is worth a load of hay;
> A swarm of bees in June
> Is worth a silver spoon;
> A swarm of bees in July
> Isn't worth a fly.

Some skepists constructed in garden walls, or even stone walls of buildings, small skep-shaped enclavities to contain their beehives and protect them from the elements. In some areas, Poland, for example, skep-shaped hollows were cut into living trees, for the protection both from weather and marauding animals. In medieval Poland the penalty for molesting another man's hives was the sudden death terrible.

If you wish, for reasons sentimental or picturesque, you may still buy such a skep (made in Holland) from certain bee supply houses. They could actually be used, but most states have laws prohibiting them now, since they do not lend themselves to inspection for disease.

The honeybee herself is perhaps the most advanced social animal going; the human race is not even in the running. Not happiest, I should think (although that too is mootable), but certainly the most advanced. The first thing one should know about this society is that it consists of a vast majority of sexually incomplete females, one sexually complete female who is the mother of them all, and varying numbers of sexually mature males called *drones*. The drones are called on only rarely for their sexual service, and that to a queen he meets out flying around (the hive's matriarch mated for life before she settled down to become the mother of them all). Each of these "three sexes" has its unique life cycle and function in the colony.

Although there are three kinds of bees in the colony, only two kinds of eggs are laid, male and female. The female egg, which the queen lays, may, depending upon treatment and handling, end up either as a worker or a queen. The vast majority, of course, become workers, the backbone and mainstay of the colony. If the female egg is placed in a normal cell, given routine care, in about twenty-one days it will emerge as a worker bee, female, but sexually imperfect; *imperfect,* not incomplete. Under certain circumstances she is capable of laying eggs, but the offspring will invariably be male; her eggs are produced by a process known as parthenogenesis (having no fathers) and can hatch only into drones.

If a female egg is placed in a larger cell, an appendage to the comb resembling a peanut both in size and configuration, and subsequently given special food and attention, she will hatch into a sexually complete female, a queen. During the first few days of its life, each larva, regardless of its ultimate destiny, is fed a particular food, rich in nutrients and hormones which the workers secrete from glands in their heads. This food is called "royal jelly," and is crucial to the first stages of larval development. After about five days, however, the diet of the larva is changed to the coarser and (from the bees' point of view) vastly more economical one sometimes called "bee bread," a mixture of pollen, honey, and certain salivary secretions. The large quantities of pollen collected by bees all goes into food for the larva, and has no part in the manufacture of honey. If, however, the egg has been "elected" to become a queen, the diet remains royal jelly throughout the larval stage. The completed insect "hatches" from its chrysalis a sexually complete female—a queen—at about fifteen rather than twenty-one days.

The drone hatches only from a male egg. The egg must be laid in an enlarged cell, placed along the edge of a regular brood comb, or in a location where the regular sizing of the cells has been accidentally displaced. While the egg can be laid by a worker bee, more usually it is produced by the queen who may, apparently by choice, lay a male egg herself. How this choice is realized in the organs of the queen is unknown, although speculation suggests that the position of the queen's abdomen in reaching into the larger opening of the drone cell mechanically prevents the release of semen, so the egg passes through the oviduct unfertilized. Thus the male eggs laid by the queen are like worker eggs, also produced by parthenogenesis.

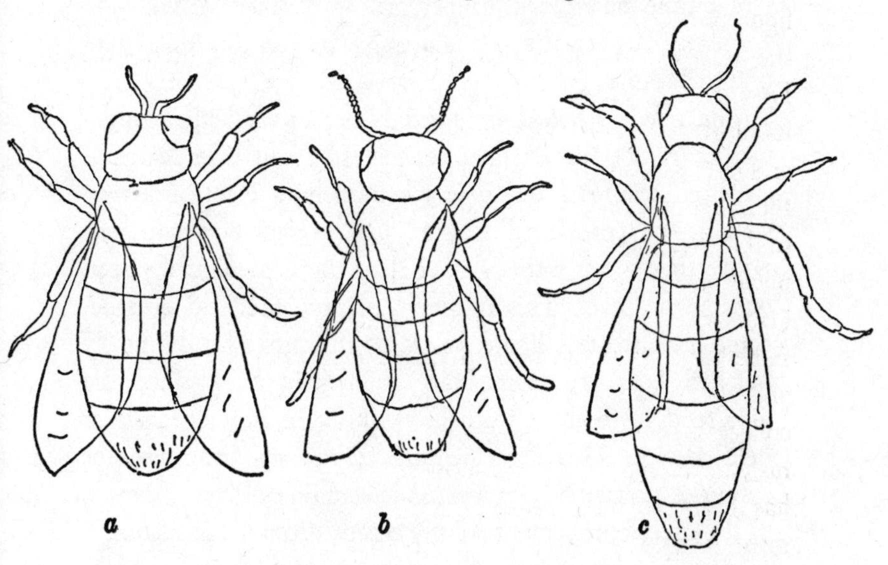

Scaled illustration of bees a. *Drone. Note the thick, stocky body and large eyes* b. *Worker* c. *Queen. The abdomen of a virgin or non-laying queen is more slender than illustrated.*

The male bee, the drone, emerges after about twenty-four days. He is a handsome insect, much larger than either the worker or queen, and true to the connotations attached to his name imperious, demanding, and utterly non-productive. He cannot forage for himself, and often will die before he will deign to even feed himself refined honey from a cell, being used to having it fed to him by the obliging workers. He eats honey only; refined honey. No nectar, no pollen. He lives for one purpose only, to service a queen. Although a large insect and given to blustery buzzing, the drone is without a sting. His posterior is engineered to entirely different functions. Despite superficial appearances, it will subsequently be seen that his is a tragic role in the life system of the beehive.

The newly hatched—*emerged* is perhaps a better word—worker is not immediately capable of flight. Her external skeleton requires a few days for hardening before she may actually be considered mature. On examination you will notice that she is rather darker than the bees normally seen outside the hives. This is simply because the field bees become bleached to a lighter shade by the sun. Also the newly hatched bees are more hairy than field bees, since the hair tends to become worn away through the exigencies of field service. This latter feature of the maturing bee for centuries prevented a proper understanding of the life cycle of the hive. Until comparatively recent times it was assumed, on human analogy, that as they grew older bees grew more hair; therefore the elder bees, pitifully denuded of their covering, were assumed to be young bees which had not yet grown their "beards."

From the time she is hatched, each worker bee goes through a virtually identical sequence of employment in the hive. The very first, and most menial perhaps, is being a janitor, and cleaning and renovating brood cells. The queen will not lay an egg in a cell which is not immaculate. All traces of previous occupancy must be removed before it is fit for its next inhabitant. Although the lines of division are not precise or clear cut, the next stage of the young bee's occupation is feeding larva. Almost simultaneously she begins to secrete wax, stimulated no doubt by the food she ingests in the process of feeding the larva.

Wax is produced by glands located between the rings of the abdomen, on either side, and the wax emerges as small, almost microscopic, discs. These the bee retrieves with its mandibles, and by chewing and adding various salivary secretions makes pliable and cohesive as she works them into whatever structure is abuilding. Working free, that is, without a wax foundation as a guide and support, bees suspend themselves in a curious loop, a necklace it is often called, working the wax and producing a finished comb. It is from the shape of this necklace that honeycomb formed free gets its characteristic elliptical shape. Certain English beekeepers use frames which exactly imitate the shape of these free-form combs, to what effect or advantage I do not know.

It is not until she is perhaps a week or more old that a bee makes her first flights. Her external skeleton is now completely hardened, her wings flat and crisp, and in all respects but employment she has become a mature bee. For several days, every sunny afternoon she and a collection of others in the same stage of development will make "training" or "orientation" flights. You will see them in a group, sometimes almost a cloud, flying in a small area back and forth in front of the hive. If ever in her life a worker bee indulges in something for the sheer sensual pleasure of it, it must be these orientation flights. After about twenty or thirty minutes, the cloud of bees will disappear as quickly as they appeared, returning to their labors inside the hive. The beekeeper should learn to recognize these characteristic flights, and distinguish them from an attack on the hive by robber bees. These orientation flights occur almost invariably in the afternoon, and last but a short while. Robbing can occur any time during the day, and proceed almost continuously. Watching the actions of the bee will show you the characteristic patterns of these orientation flights.

If you will single out a particular bee and watch her closely, you will see that she is weaving back and forth in front of the hive, fixing the hive location in her memory system. Gradually she may begin to wheel further and further from the main group, flying in widening loops and approaching the hive from different angles. This is the orientation part of the orientation flight. The homing "instinct" of the bee is almost entirely a learned visual orientation. She locates the entrance of the hive in terms of its relationship to other objects. Working in the field, she locates the hive, as well as the areas in which there is a nectar source, by a very accurate kind of geometry relative to the source of light. So much is the bee dependent on visual orientation to locate her home that moving the hive a matter of a few feet may so hopelessly confuse her that she will never find the way to the entrance. It is easier to move a hive five miles than ten feet. Moved a great distance, a bee emerges from the hive to find a new system of light angles and no familiar landmarks, so is obliged to completely restudy the situation and reorient herself to the exact location of the hive. The range of a field bee is perhaps two or three miles (usually less, unless there is a serious dearth of floral pasturage). If the colony is moved a distance beyond the radius of the bees' normal flight range, she will not stumble on old landmarks which stimulate the previous homing system, and thus be caused to lose her bearings (and her home).

Therefore, when it is necessary to move a hive a short distance, it is best to move it no more than a foot or so per day until it arrives where you want it. It also helps if the movement is not made during a heavily active period, because at such times the bees are less apt to be at their observant best. Setting up an obstacle, such as a pane of glass or even a heap of brush or weeds near the entrance, helps call the bees' attention to the fact that something has changed and encourages them to program themselves for a new orientation for the return.

Although each bee has the location of the hive fixed in its memory system, it appears that when a swarm takes off from a hive some sort of amnesiatic effect erases all previous orientations. If the queen, for example, is unable to fly and drops to the ground, the swarm will quickly perceive her loss and return to the hive. If, on the other hand, you capture the swarm and hive it, those few bees you do not manage to collect and are left behind where the cluster situated will never find their way back to the home colony. It also appears that certain kinds of drugs or chemicals may also have the effect of temporarily or permanently erasing their memories.

The young bees who are graduating into their last days of housekeeping duties occupy a unique position in the hive's social organization, at least in the opinion of many authorities.

The question is often asked respecting the complex social structure of the colony, who runs things and who decides what should be done and when. The evidence indicates the governing body to be this group of young bees, completely mature, but just before they take to the fields. The name of *control bee* has been suggested for those in this stage, and although they continue their normal duty, part of their service can almost be said to be supervisory. These are the bees which precipitate noisily and aggressively out of an ill-tempered hive when you remove the lid. If you get stung occasionally in handling your hives, most likely it will be a control bee who does the business. When the hive decides to swarm, it is most likely these control bees who make the decision, and it is they who decree that queen cells be prepared for swarming. Probably it is also these bees who keep tabs on the productivity of the queen and decide if her egg-laying capacity is failing, and if she must be superseded with a new queen. When a swarm takes off from the hive to establish a new colony elsewhere, the numbers of the swarming bees are all young bees, almost exclusively control bees or pre-control bees.

In the normal sequence of events, however, the control bee will quickly graduate into a field bee, which irreversibly seals her destiny. Happily, insofar as such matters justify observation or conjecture, the occupation of the field bee seems to be "considered" by bees universally as the prestige occupation of the hive. Anyway, at ten to twenty days, perhaps later, depending on weather and season, the young bee joins the field force. Although much of the work and occupation of the bee is instinctive, a surprising amount is learned or perfected, even dependent upon an individual bee's experience and ingenuity. When the young bee first begins to collect pollen, she will stow it awkwardly and inexpertly in her "pollen baskets," the hairy, cupping conveniences on the rear legs into which she packs this major item of bee commerce. Much of the pollen is trapped by the hairs of her body, and, hovering in the air, she combs it off herself and packs it for transport. Watching the front of the hive you will frequently notice young bees returning still liberally dusted with pollen; it takes a while for them to get the hang of how to comb it off and stow it away. The newly graduated field worker will have to learn to balance her load, how to load the baskets for maximum efficiency, and even how to fly with that weight. Watching the alighting board in front of your hives you will learn to recognize the neophytes by the tiny, ill-packed loads they return with. Inside the hive, the loaded worker goes directly to the appropriate area, and young bees appear immediately to help unload and stow the pollen. The bee will rest a short while before making another flight, eat a snack, even appearing to do the equivalence of gossiping with the other bees working on the internal economy of the hive.

Aided by instinct, bees also learn technique and finesse in gathering nectar. They learn, rather than "know," what flowers yield nectar, even to knowing what are the most favorable hours for collecting nectar from a given species. If the nectar flow is abundant and extended, given bees become specialists, even to the extent of ignoring all blossoms other than that which they have learned to understand and prefer.

The ability of bees to communicate with astonishing precision the location of foodstuff had been one of the marvels of the insect world. Dr. Karl von Frisch and his students have proven almost conclusively that bees communicate this information to each other by a kind of a "dance," a whirling gyration which, through varying speed and direction, tells both the direction and the distance to the food. Presumably the crucial element, direction, is "stated" in terms of the angle of the sun, communicated in part by the angle of the dancer's body. The distance is expressed by duration of the dance, and speed apparently says something about the quantity and quality of the source. You will not watch the front of your hive long during a busy season before you recognize a performance of the dance, with other bees standing about watching the dancer and conversing with her with their antennae. The dance is also performed inside the hive, on the frames, so presumably the contact with the antenna is as important as visual observation of the dance. This is a fascinating topic, but it is best considered in its more complete form in the technical publications, particularly, of course, the works of Von Frisch himself.

The life of the queen bee is utterly different from the life cycle of the worker. Care and attention is lavished on her from the time her egg is selected until her own life of egg laying terminates. Typically there are two kinds of queen cells made, the "swarm" cell and the "supersedure" cell. When the swarming season is toward, or when for reasons obvious or mysterious the bees conclude swarming would be a good idea, the hive bees will make a number of swarm cells, often a dozen or more. Usually these swarm cells will be made on the lower edge of the comb, and

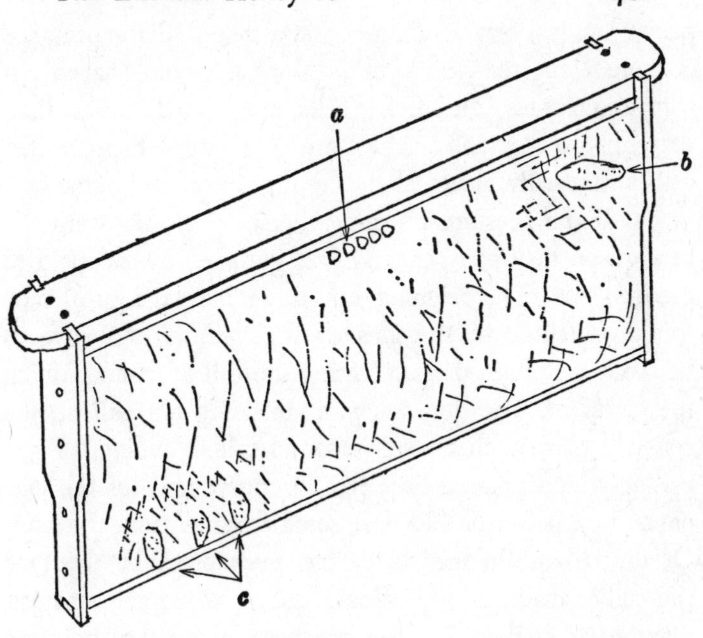

Frame of brood, showing drone cells and two kinds of queen cells (a supersedure cell and three swarm cells) a. *Drone cells* b. *Supersedure cells* c. *Swarm cells*

are easily distinguished from the regular brood cell by their vertical, peanutlike appearance. Supersedure cells, on the other hand, may be made at any time, but are constructed on the impulse of the bees replacing their existing queen because the fertility of their old queen is failing. Supersedure cells are commonly made in the center of the comb, or placed in the upper quarter of the frame, often horizontal to the top bar of the frame. The bees may disassemble a number of regular brood cells to make room for this one large cell, and the cell itself is discernibly larger and more painstakingly prepared than the swarm cell. Its appearance is more elongated and cylindrical than peanut-shaped. In comparison note the drone cells are domed rather than flat, like worker cells. They are also larger than worker cells and usually placed along the top row of the frame of a brood. If the bees are planning supersedure, they want the best queen they can get, and will go to no end of pain to produce one. The reigning queen, in her own turn, may think very little of the idea of supersedure, and have to be forcibly restrained from tearing the cell apart and killing the occupant. The queen kept from doing her will on the cell will make a characteristic sound called "piping," a surprisingly loud noise, a little like the busy signal of the telephone which can be heard at some distance from the hive. Or the old queen may be feeble, even dead, by the time the cell is made. A supersedure queen, when she emerges, may be larger than a queen produced under the swarming impulse. The supersedure queen is protected by the workers from attack by the old queen, although it sometimes seems she may have seen the handwriting on the wall and offers no objections. At any rate the old queen perishes and is cast out.

Whatever her ultimate destiny, the first necessary act of the newly hatched queen is to mate. She is an elegant little thing, a virgin queen, slender, quick, and active. Ordinarily she will probably remain in the hive three to five days before attempting her first flight. When she leaves the hive on her mating flight she flies rather low until she is some distance from the hive, most likely to escape from "domestic" drones and thus to discourage inbreeding. Drones will never mate with a queen flying below a certain height. When she has got her proper distance, the queen then seeks altitude and emits a scent which attracts drones from considerable distance, who have been cruising around to detect and track home precisely that scent. Mating always occurs in flight, about fifty feet up, as has been determined by tethering queens from balloons by minuscule threads, suspended at various altitudes.

It used to be thought that a queen mated only once, with one male, but the opinion now inclines to believe that she may mate with two or three, even taking more than one mating flight before she settles down to egg laying. If the queen fails to meet drones on her initial mating flights, she will fly out again daily, but if she is not mated by the time she is about two weeks old, she will never mate, and remain barren.

The actual mating is a rather baroque exercise; since mating is accomplished under the precarious exigencies of flight, everything must be done thoroughly and at once. For his part, the drone literally explodes in the act of mating, leaving his sexual organs implanted within the queen's body. It is said that this "explosion" can actually be heard from the ground as a kind of snap. The pair may fall to the ground, where the drone expires; the queen quickly detaches herself and takes flight, directly back to the hive. At her leisure, the queen's body extracts the semen from the drone's implanted organs and stores it within her body for a lifetime of inseminating her eggs, then discards the remnants of the organs themselves. In laying, the queen impregnates each egg with a male cell as it passes to the oviduct, apparently releasing the sperm cell by cell for optimum economy. If by accident she does not mate on her wedding flight, or mates with an infertile drone, she would of course be capable of laying only drone eggs.

As improbable as it seems, artificial insemination is practiced by bee breeders in order to improve or maintain the breed, and many breeders offer for sale queens impregnated by artificial insemination. As may be seen from the foregoing, in the course of natural breeding the breeder has absolutely no physical control over the drone with which the queen mates. Bee breeders zealously maintain their own stocks, and attempt to make sure there will always be drones from their preferred hives, actually special drone-rearing hives, flying. Even so, the possibility always is present that the queen may mate with a tramp drone, rather than one of known and proven bloodlines. Artificial insemination attempts to eliminate this uncertainty. As one would suppose, it is a very delicate and miniaturized process, requiring a steady hand and a keen eye. There has never been a bee breeder with the nickname "thumbs."

The duration of a queen's fecundity is about five years. She is capable of laying about three thousand eggs a day, but more likely in practice will lay nearer one thousand or fifteen hundred per day. A good queen, naturally, is judged by the beekeeper as well as by her workers for her egg-laying capacities. A good queen will fill a brood comb with large oval patterns of eggs, and the resultant brood will produce a frame almost solid with young. A queen which lays only spotty patches of brood is weak or failing and should be replaced. Only a highly productive queen will sustain a hive at its highest level of production.

Although the queen may be fertile and productive for as long as five years, her effective life span is much less. Beekeepers differ in their practice, but most serious apiarists maintain a plan of systematic requeening. Given their option, to be sure, the bees themselves would most likely do it for you, but without systematic efficiency. Some beekeepers requeen every year; some plan two-year cycles, and some do it only every three years. While a two-year-old queen is most likely still extremely fecund, there is much to be said for annual requeening. It is a guarantee of a young and vital queen, at what is really a rather minor expense. It also seems to repress the urge to swarm. Many—perhaps most—large apiarists maintain their own plans of queen rearing, and work diligently through selective breeding to produce and maintain the most productive and gentlest strains of bees. While a dollar and thirty cents or so for a new queen is a relatively minor annual expense for a hobbiest, a man with three to five thousand hives would find it anything but minor.

Of course the obvious question suggests itself here: why not let nature take its course and let the bees requeen themselves when it seems appropriate to them? Several reasons argue against this. One is that the declining efficiency of the queen may result in a hive existing below standard for a year or two before being revitalized by a vigorous new queen; or worse, it may weaken to the point that it is incapable of requeening itself, and perish. Then, too, there is the obvious difference in strains of bees. Some bees, even of the same variety, are more productive than others, winter better, or are better tempered. Certainly for the amateur, it is best to buy the choice strains of queens from professional breeders to get the advantages of the best possible strains of bees. If you are a hobbiest attempting to maintain a specific strain, such as Caucasian or Carnolian, there is much more likelihood that your queens will mate with a tramp drone than one from your own hives, in which case you have effectively lost your strain forever, until you requeen. Good queens are inexpensive and you will never go wrong in buying them.

When you order a queen, either with a packaged swarm or for requeening, specify that she be marked; it will add only about ten cents to her price, and a marked queen is much easier to locate in the frames than an undistinguished one. Also there is a color code which will tell you, if you do not requeen all of your swarms at the same time, when a given queen has been introduced. It is possible to order a queen with a wing clipped (or you may do it yourself) which keeps her from flying, and also helps identify her after she is in the hive. There are advantages in swarm control in having your queens clipped (however, it will not in itself prevent swarming), but it is perhaps finally a matter of preference. Clipping, by the way, in no way impairs the function or efficiency of the queen. If you do it yourself, handle the queen with great care, picking her up only by the thorax, never by the abdomen. Mishandling a queen may impair or destroy her egg-laying capacity.

The life cycle of the drone is a rather melancholy one. His one function in life is to service a queen, and if against dismally slender odds he does in fact mate, he dies, almost instantly. Nevertheless, when a queen makes her wedding flight it is so important that she rapidly be found by a drone that each hive devotes considerable pains to maintaining a group of them. When a queen leaves the hive, she is vulnerable to wind and weather, attack by birds, and innumerable other hazards that minute flesh is heir to. Her scent alone assures that she will be found in the wide, wide sky by a male, and there is little time to get this done. On every good day of the late spring and through the summer, therefore, drones must be in the air on station and waiting. Evidence confirms a tradition that drones tend to congregate in specific localities, waiting for a queen to appear. This seems equivalent to an insect version of the corner drugstore. At any rate, such "collecting places" seem to be sought out by both drones and queens. This subject, like a great many others respecting the natural history of bees, needs further study.

A thriving hive will usually always maintain a few drone cells. From the beekeeper's point of view, the fewer drones the better. The cells are wasted space, and rearing the drone brood and caring for the mature drones takes time and productivity from the worker bees, which could more profitably be bestowed elsewhere.

A drone egg may be laid either by a worker or a queen. Occasionally, because of a failing queen, or a queen being lost altogether, a colony may turn into drone layers. This condition usually may be noticed quickly if the hives are examined regularly, because of the number of drone cells—whole frames of them—and the absence or scarcity of worker brood. Unfortunately this is a very hard condition to correct, as at this juncture it is very hard to induce the hive to accept a new queen.

The life of the drone is relatively long, unless he fulfills his destiny by mating with a queen. In spite of his being a drag on the domestic economy of the hive, he is suffered gladly, tolerated and even pampered by all. Drones seem to be accepted in any hive, regardless of where they actually reside, and may request and get a handout in any hive, where their sisters and mothers would be regarded as foreigners or bandits and assassinated on the spot. This universal hospitality which is extended to drones likely results from their range; while this matter is not known exactly, it seems that the haunts of the drone may extend much beyond the flight range of the ordinary field bee; some think it may be six or eight miles. During a time of want, however, a scarcity of nectar during the summer, or at the onset of autumn—the tolerance for drones ends. With utter lack of sentiment, drones are summarily excluded from the hive and denied food by their sister/mothers. Since the drones are incapable of feeding themselves, they quickly die. It is absolutely characteristic of autumn in the apiary to see crowds of rejected drones on the alighting board in front of each hive, sternly being denied entrance. And as the autumn deepens, in front of the hives and generally around the bee yard will be seen the dead husks of drones, finally having succumbed to starvation and the weather.

4

Hiving a Packaged Swarm

BEFORE YOUR BEES ARRIVE you should have selected the physical location where you intend to keep your hives permanently. A location which will provide a certain amount of shade, without a heavy canopy of overhead branches, is ideal. The hives should get the morning sun, if possible; as a rule bees are rather late risers. They do not begin to work outside much until the sun begins to warm the hive. Therefore, to discourage sloth, the hive should catch the sun as soon as possible. No beekeeper wants to have to get up, go out in the apiary, and wake up the bees himself. For some of the same reasons the hives should not be situated in any sort of hollow or declivity of the ground which might tend to collect or hold cold air. This fact is important both in encouraging early activity of the colony during the working season, as well as being a consideration for successful wintering of the colony. As much as possible the location should also protect the colonies from the prevailing winds; a fence, if no other protection is available, can be erected, if nothing more than a bamboo screen.

Generally speaking, then, a desirable location for the colony should have an unobstructed pathway to and from the hive, protection from the overhead sun if possible, and access to the early morning sun. If the hives must be situated close to a dwelling, or near a route frequently traveled by humans, place the hives so the entrances face *away* from the residence or line of travel. This will reduce the number of bees which kill themselves stinging people.

Hives should not be placed closer together than a minimum of 18 inches. A great many bee books offer numerous suggestions for the layout and organization of the hives in a bee yard, many oddly reminiscent of Sir Thomas Browne's tedious treatise on the quincunx arrangement of Roman orchards. The reason for separating the hives is to prevent "drifting," the accidental displacement of bees from one hive to another through the natural absent-mindedness of hurried worker bees during a busy season, aided by the effect of the prevailing wind. The proprietor of a modest-sized apiary can forego elaborately planned arrangements.

As much for the protection of the wood against decay as anything, the hives should not be placed directly on the ground. Some beekeepers even establish permanent concrete pedestals for their hives, or use pre-cast concrete stands which are theoretically portable. Prefabricated stands of pine or cyprus may be ordered, which are both attractive and inexpensive. [Most hive stands are so simple, a beekeeper may prefer to make his own.] Actually it is sufficient that the bottom board simply be protected from contact with the ground (but the stand should not allow a draft of circulating air under the hive). Some sort of alighting board should

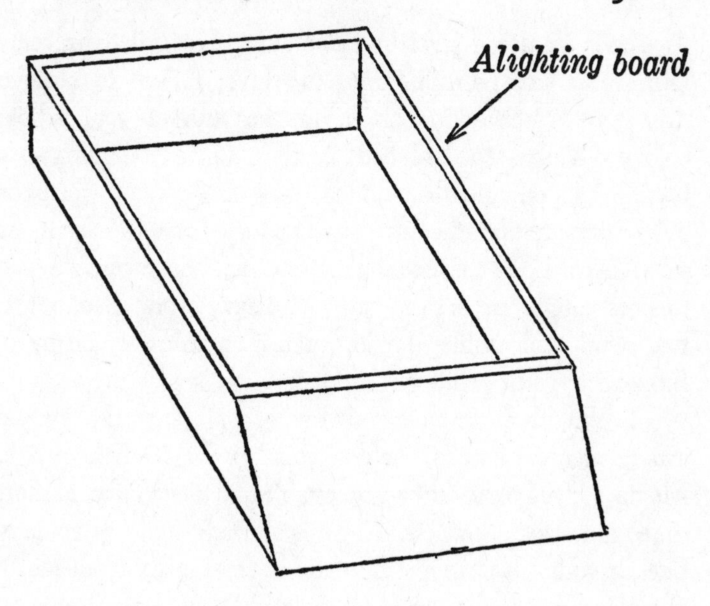

Hive stand

be provided in front of each hive, which may be no more than a flat piece of plank the width of the hive, ten or twelve inches wide, and at the same elevation of the hive entrance. This makes a convenient landing area for the returning bee, which is none too steady on its wings anyway when it comes in with a heavy load, particularly in a light cross wind. The alighting board also makes observation of certain hive activities particularly convenient, an appeal which no hobby beekeeper can resist.

Perhaps the most important thing to consider when you physically place your hive into position is to make certain that you level it until there is a slight pitch forward. Don't "eye-ball" it; use a level to make sure. This pitch prevents rain water from running into the hive; failure to observe this simple precaution can mean the ruin of your whole colony during a spell of bad weather. Once your colony is hived and permanently situated, take care to cut weeds or grass from around the hive, particularly from the front, as such obstructions are a serious annoyance and inconvenience to bees caught up in the fever of heavy work. Bees relish hard work and, unlike humans, detest anything which interferes with getting the work done.

You should either have your hives in place, or the specific arrangements planned, before your packaged bees arrive. Moving hives, once they are established (even for a short time) is a tricky business. There is also advantage to actually having your hives in position, or at least placed outside, a few days before you plan to establish your bees simply to air them and make certain all paint fumes, etc., have been cleared away.

To the considerable consternation of your mailman, the day arrives when your packaged bees are delivered. Quite likely instructions for their immediate care will accompany them. The sugar syrup shipped with the bees in the carton should have sufficed to keep the bees in good condition, even if they have been shipped clear across the continent. (In my locale, the state of Washington, bees are commonly ordered from Texas or Georgia, and arrive fat and sassy.) As soon as your bees are delivered, however, you should give them supplemental feeding of additional syrup, as much as they will take. They will appreciate it, and it will give you a sense of hospitality. Mix a solution of two parts

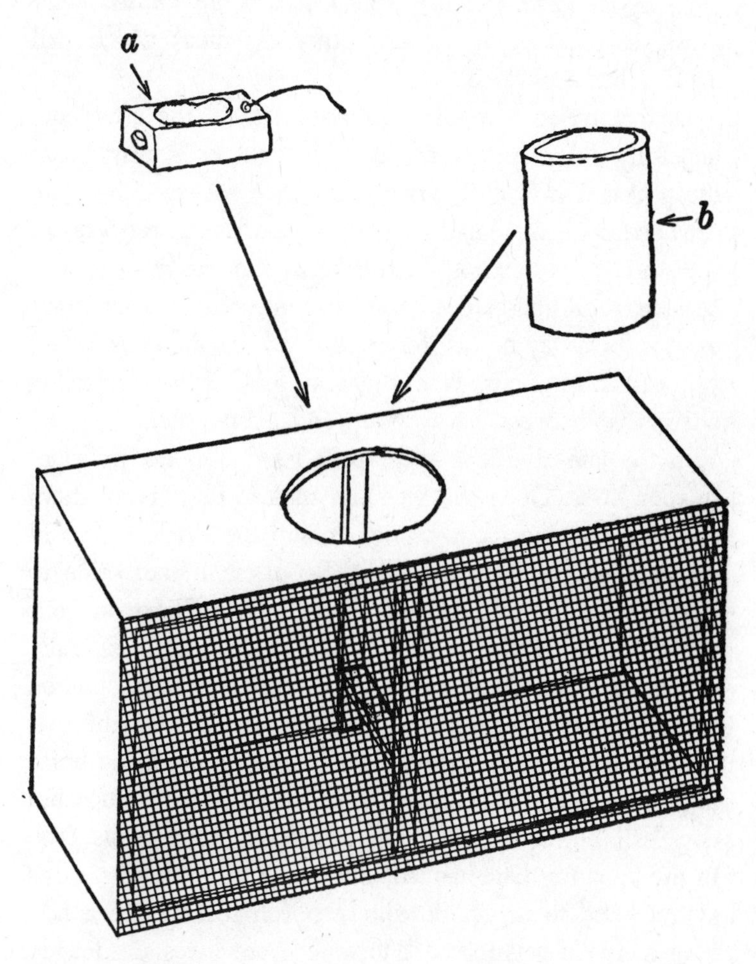

Screened shipping cage a. *Queen cage* b. *Syrup can to feed bees in transit*

sugar with one part warm water, and spread on the wire mesh of the cage; you may spread it with an *unused* paint brush, sprinkle it, or use any other expedient which will get the business done.

Do not try to hive the bees until late in the afternoon, hopefully an afternoon which is not wet or stormy. Keep the packages in a cool, dry place (more packaged bees are damaged by heat than by cold). If you use insecticides in your house, pest strips or sprays, *by no means* keep the bees in your house, even for a minute. If it is necessary to wait for a day or two for good weather conditions before you can hive them, feed them at least daily, preferably alternating between sugar syrup and plain water.

In the late afternoon—perhaps 4 P.M.—take the packages to your hives. Open the hive and remove four frames from one side (leaving room enough to slip the whole carton in which the bees were shipped). Set these frames aside to be returned to the hive several days later. Have on your person pliers, knife, a six- or eight-inch length of thin, flexible wire, and a tack or two. If you feel timid, put on your veil, but the bees will be very docile. Everything is on your side; the bees are young, disoriented, with no home to protect, and full of syrup. Wear gloves if you must, but they will be an impediment, and mostly superfluous. You might be stung bare-handed, but this is a good time to be getting used to it; and it will keep you feeling like a bee-keeper after it gets too dark to watch your hives any longer. Fire up your smoker; a good fuel is a rolled up piece of burlap, old toweling, or the like (do not burn wool or synthetics). Some beekeepers use dry corncobs or leaves;

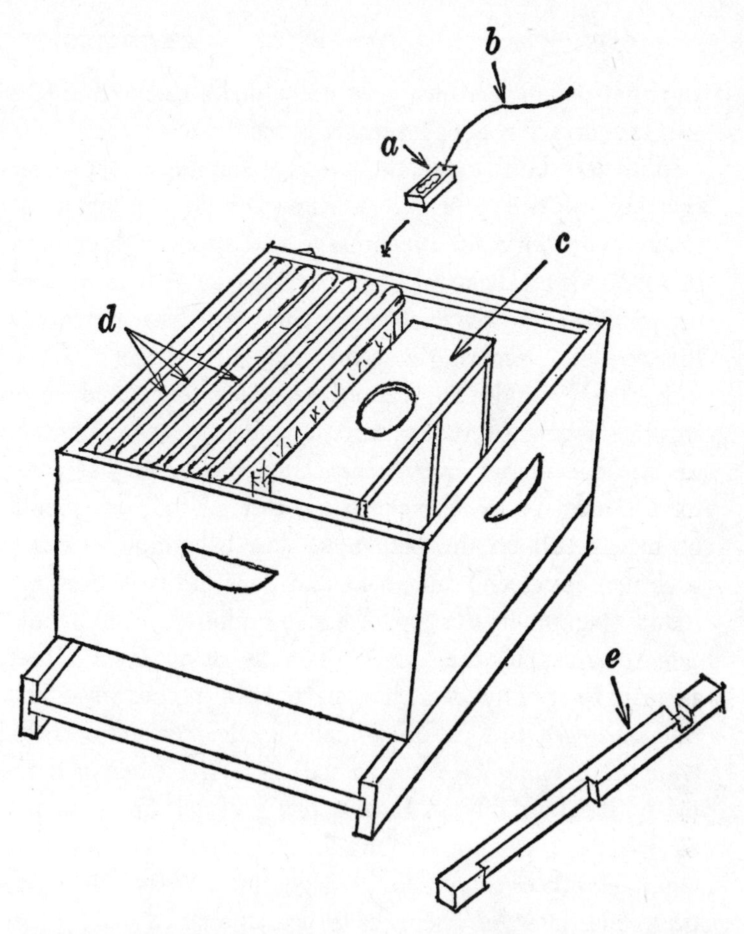

Hiving a packaged swarm of bees a. *Queen cage* b. *Wire attached to queen cage by thumb tack* c. *Bees' shipping cage* d. *Frames filled with foundation (or drawn comb, if you have it)* e. *Entrance block, with variable-sized entryways. Fit it to the hive entrance with the smallest opening in position. After about three weeks, readjust the block to put the larger opening in position. It should be removed altogether after about six weeks. It is replaced in late fall, with the larger opening (which allows the use of a Boardman Feeder) in position.*

anything non-toxic which puts out a dense cloud of smoke and smolders for a long time is serviceable.

With your knife loosen the tacks holding the small circular cover in the center of the package, but do not remove it immediately. First lift the package and strike it sharply on the ground to shake all the bees to the bottom of the carton. Now quickly remove the cover you have previously loosened and remove the syrup can. The queen will be immediately inside the carton, separately contained in a small semi-screened wooden cage suspended by a wire. Quickly lift the queen cage out of the carton and place her aside for the moment. Now pick up the carton (the bees should be mostly still on the bottom, or just beginning to climb back up the screen) and shake and pour the bees from the round opening in the top of the shipping cage on to the center frames of the hive. When as many of the bees as may be readily detatched have been poured over the frames, lower the carton containing the remaining bees, with the opening face up, in the space left when you removed the four frames. Blow a few puffs of smoke in the air over the hive (but not into the hive itself) to discourage flight. Most bees shipped in packages, by the way, are young bees, and most of them have actually flown very little; there is little likelihood of their trying to fly excessively when they are poured from the carton.

Now take the queen cage, and with the point of your knife remove the small cork protecting the candy core leading to the queen herself. Note well: there is a cork in each end, the one placed in sideways, the other "normally." The cork which is on its side opens a direct passageway to the

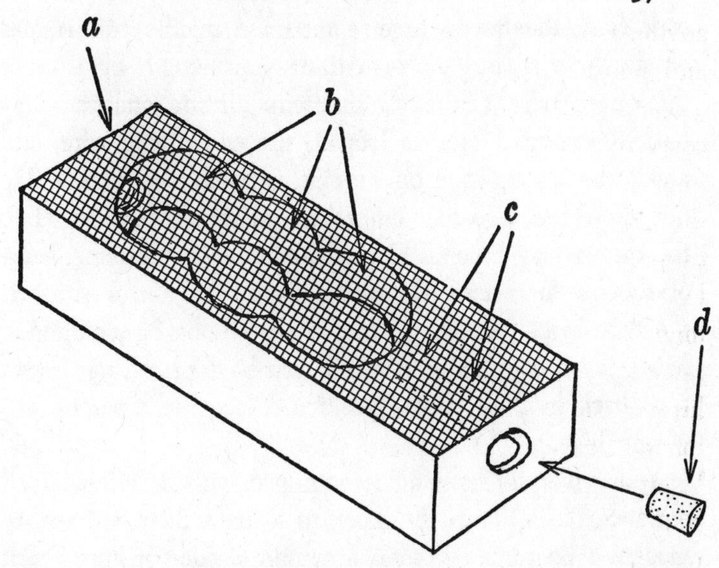

Queen cage a. *Direct access channel to queen chamber, tightly closed by a cork positioned sideways* b. *Chamber in which the queen and her attendants are free to move* c. *Channel filled with candy* d. *Cork inserted normally, which is removed when the queen cage is placed in the hive, to allow the eventual freeing of the queen. The queen cage is covered on the open side with fine wire screen. Do not remove this screen.*

queen; the other protects a core of candy, which the workers will eat away in the course of a day or so, thereby becoming her familiars before she is actually released. This latter is the cork to remove, the cork which is inserted the "right" way. If you open the direct passage, the queen escapes immediately, and runs a severe danger of being killed by the bees of the new colony before they become settled and accustomed to her. Having removed the proper cork from the queen cage, spread the middle two frames of the hive slightly apart. Attach your length of wire to the queen cage (although the wire already attached may be long enough), from a loop in the end of the wire, and lower the queen cage on its side between the frames. Do not allow the cage to hang from the wire, the channel to the queen may become blocked and prevent the emergence of the queen. Secure the queen cage with a tack through the wire loop, fastening it to the top of one of the frames. Move the frames between which you have placed the queen cage back in their proper position. Place the cover on top of the hive.

Your hive will have come equipped with a variable-sided entrance block. Turn the block to a size which will accommodate a Boardman Feeder—a simple wooden mount which accommodates an inverted mason jar with a perforated lid—leaving a small passageway for the traffic of the bees. Position the feeder so it will be beneath the frames, not the empty space occupied by the shipping cage. Next fill a quart jar with sugar syrup (or have it filled in advance), screw on the lid, and place the jar in its inverted position in the feeder. It all sounds very simple and neat to describe it, but the first time you hive a swarm you may feel flushes of panic, and end up sure you've botched the whole thing completely. In spite of the fact that the bees are young, some will be flying around, in what might seem to a beginner a voracious crowd. Hopefully most of them will find their way inside. The rest will be lost, and, as Dr. Johnson would say, "that's an end on't."

You will probably hive your second colony more gracefully.

Now begins the most anxious and crucial part of establishing your swarm. Except for refilling the feeder jar (no definite time schedule can be given; just watch it and refill when it's empty), for the next seven days do not open, examine, or molest the hive in any way. This restraint is highly important. It takes that long for the bees to become fully familiar with their new surroundings (although they will begin to draw comb immediately), and completely accept the queen. Actually the bees will be pleased and gratified, within their capacities, with the new, clean, roomy home that smells of beeswax which has been provided them after their constrained sojourn. But until they are fully established, if they are examined they may resort to a practice called "balling the queen." This activity is not completely understood; it may be intended to protect the queen, it may be a neurotic startle reaction, or it may result from unexplained impulses to kill the queen. At any rate, this latter is the usual result: the queen will be killed. Worse, probably some time will pass before you recognize that the queen is gone, perhaps so long that it will be too late to requeen successfully.

During the seven days of waiting before you can open the hive, you will probably spend as much time as you can watching the entrance. Most likely, weather favorable, on the day following the introduction of the new swarm to the hive you will see bees returning with conspicuous lumps of brightly colored pollen on each hind leg. That is a good sign, a sign that probably the queen has already been released and accepted, or at least that the hive considers itself to have a queen, if the actual release is not complete.

Also keep alert for less sanguine symptoms; sometimes the queen may not have been accepted, in which case you may sometimes find her physical remains in front of the hive, alas; a trim, attractive insect somewhat larger than a worker bee. Do not be unduly alarmed, however, if the bees seem to be carrying out the bodies of what appears to be a large number of dead bees. Many will have been injured in transit, and a hive will remove dead bees as soon as possible.

After seven days, the hive should be inspected, as briefly and undisruptively as possible. Probably it will not be necessary on this inspection to use smoke, if your investigation works out to occur on a warm afternoon when the bees are working up to capacity. First remove the empty bee carton and replace the empty frames you had removed previously. Sometimes there will be a number of bees still in the shipping cage; occasionally they will even be found to have hung comb in its interior. If there are bees in or on it, leave the cage near the front of the hive. They will all return to the hive before night. Lift one of the frames on which the bees have been working, from what appears to be the center of the cluster (i.e. the mass of bees working on the frames), and one in which the bees are drawing brood comb. Look carefully in cells which have been fully drawn, and you should see either eggs or tiny larvae. If you find neither, check the other frames; if the results are still negative, the queen is either dead or not laying, most likely the former. Order a queen immediately, to be sent by airmail. There is still ample time to salvage the hive.

Most likely, however, everything will be found to be running smoothly on this first inspection. The bees will probably be quite easy to handle. (There aren't really too many of them, and you will find yourself working with confidence. You may even begin to fancy yourself something of an expert.) Probably you will be astonished at how quickly the bees have made themselves at home and turned the box you had prepared for them into their own beehive, something which is entirely theirs, and only in the most mechanical ways belonging to you.

After this inspection which is given at seven days, there yet remains another fourteen most crucial days, during which the life of the colony will flicker to its lowest ebb. The reason these days are so crucial is that it takes a total of twenty-one days from the time an egg is laid until the new bee hatches. Remember, for that length of time your hive population will consist solely of the original bees which you received in the package. While package bees are almost all young, nevertheless they will have had the full burden of establishing a new hive, including duties usually spread over several age groups, rather than just one. Also you must consider the dreary fact that even barring accidents, the life of a field bee during a good working season is deplorably short. The first brood of your newly established colonies, then, will not hatch until a minimum of twenty-one days after installing the swarm; most likely a couple of days longer, since probably the queen will not have been released and begin to lay until at least the passage of a day or two. This period, then—from twenty-one to twenty-five days after installing the swarm—will find your colony at its weakest. You should therefore by no means fail to keep your new colony amply supplied with sugar syrup, throughout this time. In addition to feeding the three or so pounds of adult bees you ordered originally, plus an undetermined but increasing number of larva bees, the swarm has been using a tremendous amount of nutrient in making wax. While it is not strictly necessary, I recommend feeding pure honey rather than sugar syrup during at least part of this crucial period. It costs more, to be sure, but for wax production it is a great deal superior to syrup. A commercial beekeeper would never feed honey (unless he had a surplus of low-grade or unmarketable honey), but an amateur can afford to let his sentimentality lead him to certain lavishnesses.

With the passage of the magical twenty-one days, marvelous changes begin to take place. Although the increased number of bees inside the hive is immediately apparent when you begin your inspection, it will still be a while before the activity outside the hive is appreciably affected. For about the first fourteen days of her life the newly hatched bee will be almost exclusively an inside worker, taking care of brood, building and cleaning wax, housekeeping, and the like. But because there is now an increased number of bees inside the hive, you may now begin to do a little housekeeping of your own. At perhaps twenty-five days after installation, open the hive and move what appear to be the center two frames of the colony (i.e., the center of the cluster of bees) *one position outward*, and refill their vacancy with frames which the bees have not yet begun work on. These frames which you are repositioning should be well worked (i.e., the foundation drawn out into cells), and filled with capped or hatching brood. The purpose of this maneuver is to encourage the colony to expand its brood nest by enlarging the area upon which the queen is laying. It is important that the chamber not be enlarged more than by the width of the two frames,

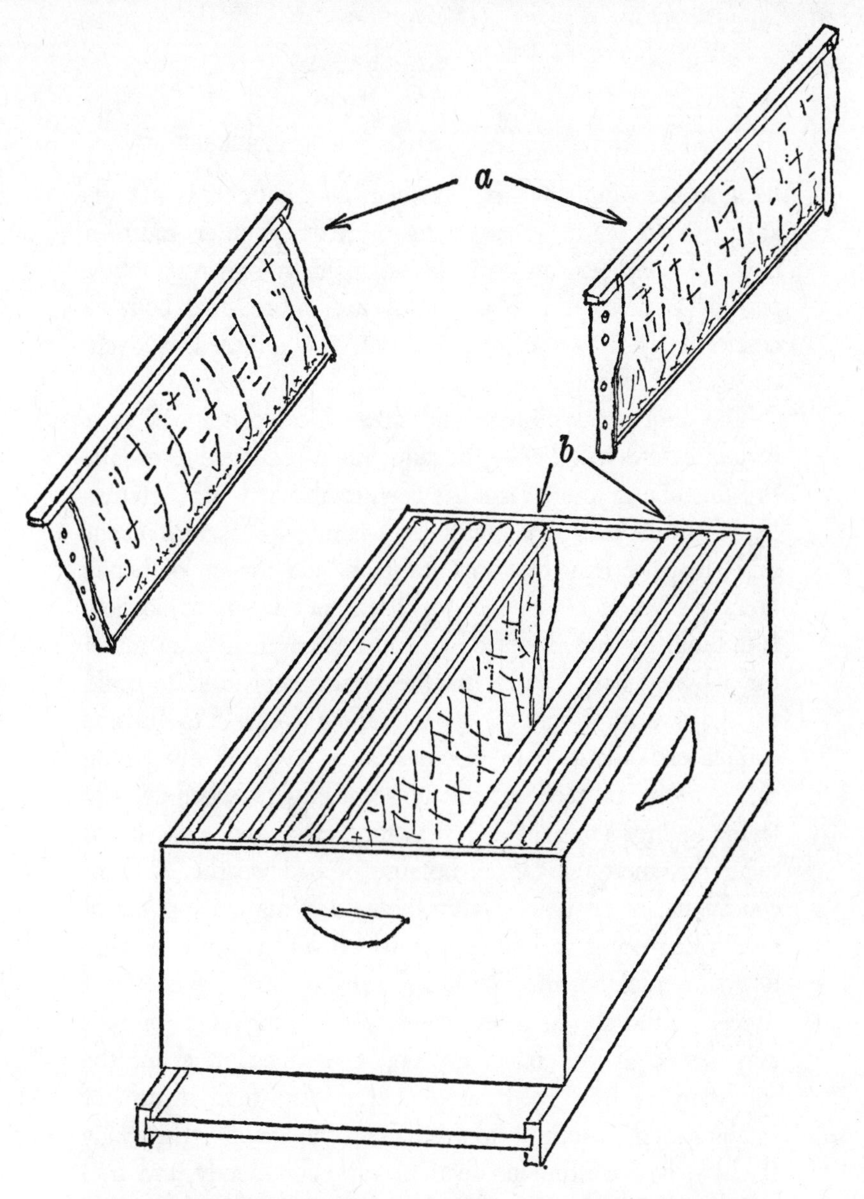

Enlarging the brood chamber or a recently hived swarm of bees. The drawn comb (a) *with brood are removed and the flanking frames* (b) *are moved together to the center, now themselves to be flanked by the frames* (a).

because the number of bees present at this point is not yet adequate to warm a larger mass. After another fourteen days the chamber can again be enlarged by moving another pair of frames; and so on, until the entire hive body is occupied by drawn comb in which the queen is actively engaged in laying.

The individual climate and other local conditions to a certain extent will dictate the subsequent sequence of events. Presumably, if your swarm has been established—say, arbitrarily, about April 1, for a northerly climate—it is now about mid-May, the fruit blossoms are gone, and the major honey flows are yet to come. What a beekeeper wants to do now is to build his swarm up to maximum strength in preparation for a honey flow; not to use the honey flow itself to build up the swarm. Therefore do not allow yourself to become complacent about your colonies now, even if everything seems to be proceeding exactly according to schedule, egg laying going along at a satisfactory rate, and new brood regularly emerging. Often, because of bad weather or local conditions of natural contrariness, a situation of actual dearth can occur in late spring, which will cause a slowing down or total cessation of brood rearing during the crucial time of build-up. In severe cases, general starvation of bees may result. If you have any doubt or question about the condition of the food supply after the first month of establishment, continue to feed them moderately. Especially if there is a continuous spell of rainy or cloudy and cool weather, you should supply food. When the real summer blossom period begins, you want a healthy, happy colony of young, ambitious bees ready to go to field for you.

5

The Vulnerable Bee

THE BEST DEFENSE AGAINST disease in your apiary is of course preventative. Do not employ used equipment if there is the least uncertainty about its origins and previous conditions of servitude. Provide preventative medication in the fall and/or spring, as described infra, and maintain conditions in your colonies that will encourage strength and resistance to infection. A strong hive is much less susceptible to contagion than one which, through various reasons, has become run down.

The most disastrous diseases for American beekeepers to contend with are those affecting the brood. A colony—or an entire apiary—may be literally wiped out in short order by such diseases. The most ravaging of all these diseases is *American foul brood*. In spite of its name, it did not originate in America; naming it seems an example of attaching to an infection the name of an unpopular foreign country, as "French pox," "Italian disease," "English complaint," etc., for syphilis in the sixteenth and seventeenth centuries; and Asian and Hong Kong flu in more recent times.

American foul brood attacks larvae in all stages of development as well as the pupae of the bees. The commonest and most decisive test for American foul brood is to check the physical consistency of the dead brood with a match stick or similar implement. (There is probably for sale somewhere a very expensive instrument for doing exactly the same thing.) Characteristically the dead larvae or pupae can be pulled out in a stringy mass, while the main substance of the dead larvae clings tightly and messily to the cell. The dead in the pupae stage are found with their tongues completely extended, sticking upward. This characteristic is distinctive of American foul brood, and is not otherwise to be found. Usually a strong, unpleasant odor, sometimes described as "gluelike," is associated with American foul brood. The odor of course is the origin of the generic name, "foul brood," applied to both American and European foul brood, although the two infections are produced by entirely different micro-organisms.

Once a colony has become afflicted with American foul brood, little can be done other than to prevent the infection spreading to healthy colonies. Its cure is possible, but scarcely practical. The bees in an infected colony should be killed by fumigation (usually cyanide, or sulphur fumes) and the hives should be burned in a pit that will permit covering the ashes with six or more inches of soil. Do not attempt to salvage any of the combs or frames of honey (after fumigating the honey will be dangerously contaminated). While some authorities contend that even a badly infected hive may be saved with the use of modern drugs, the Department of Agriculture does not concur, and by legal authority insists on disposing of the hives. Since the consequences to your colonies are so severe, it is readily apparent why one should be most vigilant in preventing his hives from becoming contaminated by used equipment, and why preventative medicine should be practiced as a matter of routine.

The disease is spread also by healthy bees robbing weaker, afflicted colonies (often nearby wild colonies whose existence you do not even suspect). It may also be introduced into your apiary from a wild swarm which you have captured, or a colony you may have obtained from a box hive or gum (the hollowed-out log hive popular until recently in the southeast), and rehived. Even having used your hive tool while working with an infected colony, it is said, may be sufficient to introduce spores of the disease into healthy colonies.

While it may be true that the disease may not be effectively cured once it gains a foothold, its prevention is reasonably sure by using sulfathiazole or Terramycin. This medication is provided in different forms (sulfa is sometimes routinely added to commercial preparations of pollen substitutes), and is advertised in the bee journals and the catalogues furnished by supply houses. Both medications are usually administered in syrup, but in some forms sulfathiazole is applied directly to the frames as a powder. Bees may be medicated either in the spring or fall. Use the dosage recommended with your particular brand. Medication should not, however, be administered within a month preceding a honey flow.

A second common disease of brood is *European foul brood.* This infection attacks the brood exclusively in the larva stage, and usually the dead larva dry into a small scale which adheres to, but is readily detached from, the side of the cell. The infection is also accompanied by an unpleasant, sour odor, but not as pronounced or characteristic as that of American foul brood. A strong colony may be able to control this infection by itself, although it may in the process be so seriously weakened that the colony may succumb from other causes, including starvation. Destruction of infected hives is not mandatory, as in the case of American foul brood, since it is possible to cure it. The infection may be prevented or controlled by spring or fall medications, like American foul brood. Commercial preparations of Terramycin are administered, either in dry form, mixed with powdered sugar, or in soluble form, mixed with syrup. The soluble form tends to lose its effectiveness rather rapidly.

A third affliction of bees is called *sacbrood.* It usually attacks the larva after the cells are capped, and unlike either American or European foul brood, usually relatively small quantities of brood are affected. It is a most serious threat when the hive has a queen of reduced egg-laying capacity. An infected hive should be requeened, to make sure of optimum brood production. The colony should be able to take care of the infection themselves; its most serious threat is to production. In areas of heavy infestation it is a common cause of poor honey yield. Antibiotics are apparently of no effect against it.

Occasionally special circumstances will bring about a condition known as *chill brood.* This is not a disease; the condition occurs because the cluster of bees is not large enough to maintain the warmth proper to the development of the brood. Usually only small numbers are affected, and these mostly in the outer frames. More often than not the beekeeper himself may be responsible for this condition—leaving frames of brood outside the hive too long during inclement weather, disturbing the coherence of the cluster during cool or cold weather, or misplacing frames of brood back into the hive and disturbing the capacity of the cluster to keep the "nest" warm. Therefore when you are "spreading" the brood nest in a newly hived colony, take care not to spread the frames beyond the capacity of the existing bees to keep it warmed. In an established colony, generally speaking you should put the frames back in the same place you took them from to prevent misplacing a frame or brood so far from the center of the cluster that it suffers neglect, allowing some of the larvae to die.

A common disease of adult bees is called *nosema.* This infection affects the digestive tract, and seems especially prevalent in northern areas, where the periods between winter cleansing flights are especially long because of weather conditions. Nosema infection may reduce the effective life span of adult bees by more than half. The organisms of the disease seem to be spread especially through contaminated water, and the availability of fresh, clean water is indicated as a preventative measure. The disease usually reveals itself by adult bees crawling around in front of the hive, unable to fly, walking awkwardly or crablike, and with distended abdomens. The presence of abnormal numbers of dead bees around the entrance of the hive, in absence of obvious causes such as accidental poisoning by insecticide, also indicates the possible presence of this disease.

Nosema infestation will not destroy the colony but will seriously cut into its productivity. Its presence may often go unnoticed, and the beekeeper may attribute a colony's poor showing to bad luck or other mysterious causes beyond his ken. The infection may be controlled by use of fumagillin, an antibiotic. Like the drugs mentioned above, it is available through the usual sources; follow the directions for the particular commercial preparation you obtain. Again, medication should be administered in the early spring, at least a month before the first honey flow from which you might expect a surplus.

An infection which is fortunately only of theoretical interest (yet) to American beekeepers is the acarine infestation, otherwise known as the Isle of Wight disease, from the name of the place where it first appeared. It is caused by a submicroscopic mite, akin to the cheese mite, which infests the breathing tubes of bees, interfering with their respiration to the point the bees are unable to fly, and eventually die—chiefly due to the accumulation of fecal matter. In the early years of this century, when it first appeared, this infection nearly obliterated British beekeeping, and did in fact effectively exterminate the native British black bee. The actual cause of the ailment went undetected for years, since the symptoms suggested infection of the alimentary tract rather than the actual cause. Once the cause of the infection was discovered, means of control were not long in being found. This mite presents a curious example of parasitic evolution; no one can guess how many years this minute creature existed on other hosts before it branched out and discovered its true destiny, with such disastrous results, among the bees of the Isle of Wight.

Besides disease and parasitic infestation, there are other more visible pests which may at times inconvenience and discommode your bees. One of the most serious and persistent of these is the wax moth. While these creatures seldom invade a strong colony, they may seriously add to the burden of a colony which is not up to strength for the varieties of reasons a hive may be debilitated. The larva of this moth consumes the wax in combs, trailing silken webs behind it until the entire comb may be tunneled—to confuse a metaphor, *honeycombed*—and rendered useless. Infection is commonly spread by using contaminated equipment, although the moths themselves may enter a hive and establish a breeding cycle. Wax moths are also especially pestiferous in attacking stored combs or foundation. Professional beekeepers commonly fumigate their supers of comb before they store them for the winter with commercial preparations effective in the control and prevention of wax moths.

Ants may also be a serious annoyance to your apiary, and may be difficult to control because the same baits that are attractive to ants may also be attractive to bees. Ideally, trail the ants to their little lair, and destroy them there without mercy. Do not use poison that the ants may track back into the hives before they succumb. Mixtures containing chlordane are generally to be considered unsafe. I have found that the small pillbox-shaped bait traps which have the poisoned substance (usually arsenic) safely inside the metal can and out of the reach of bees, etc., to be reasonably effective. If you gain control of the ants in the spring, usually they will cause no further problems.

Hives will also sometimes be invaded by yellow jackets. Most of these invaders are carnivores, hunting the bees themselves rather than seeking out the stores of honey. The best—really the only—defense against yellow jackets is in the strength of the hive itself. A strong colony will usually be able to repel the invaders, or at least to keep them under sufficient control that their depredations will be minimal.

Among the larger pests, skunks may also attack hives of bees, chiefly at night. Their interest is also in eating the bees themselves, which they induce to come out by scratching on the front of the hive. As the alarmed bees emerge, they are eaten, presumably with relish. Mice may invade the hives, especially in the fall, seeking a snug wintering place as much as the substance of the hive. Restricting the opening of the hive in the fall will prevent their invasion, although some beekeepers further make their hives proof against invasion by a series of little teeth between the bottom board and the front of the hive. If mice gain entrance, more often than not the bees will kill them promptly, and "embalm" the corpse in propolis, the waxy substance they gather so freely especially in the fall. Then there are bears; I hope you are not so unfortunate as to be afflicted by bears.

Birds of various sorts, particularly king birds, may capture the bees on the wing. While such snacking certainly doesn't do the economy of your hive any good, it is not likely to inflict crucial damage on your apiary. Other birds similarly snatch up an occasional unconsidering traveler, and there's very little you can do about it. Respecting these undesirable occurrences that fall short of disasters, the best you can hope for is that the results be moderate, and otherwise simply to learn to live with the inconveniences. Like raising oak trees, beekeeping tends to teach one to look to the long view.

6

The Seasons of the Bee

The end product sought in managing bees through their seasons is to put your bees in condition to build up to strength for the strongest honey flow rather than to build up *on* the honey flow. That is, the period of maximum production should find your colonies at maximum strength, and not be used as a period of building strength. *When* "spring" comes to your colonies depends mostly on your geographic location. If you are just beginning with package bees, you will hive them sometime early in April, feed them fairly constantly, probably well into May, and that will have been all you can do for spring build-up your first year.

If you have just wintered your first hives of bees, however, you will almost certainly want to give assistance in the spring build-up which was not possible the first year. You will want to begin providing them with some quantity of supplemental food in the very early spring even if they have ample domestic stores, if for no other reason than that to do so stimulates early brood rearing. By far the best supplement for stimulating the rearing of brood is natural pollen. If you like to tinker you can collect your own pollen—that is, you let the bees do it and then collect the pollen from them. You will need a pollen trap, which you can buy from most any bee supply catalogue. Its working principle is to force the bees to "squeeze" between wire rods placed so close together that the bee will dislodge the pollen on her legs as she works her way through. Pollen traps completely block the front entrance of the hive, so they should not remain in place more than a few days at a time or the hive will itself be deprived of food for its own brood. Also, like any encumbrance which restricts the free passage of the bees into the hive, it definitely cuts into honey production. Beekeepers who collect pollen in commercial quantities report that the bees quickly learn certain wrinkles to at least partially circumvent the trap, particularly the expedient of carrying smaller loads of pollen and thus reducing the likelihood that the loads will be dislodged by the wire rods. Somehow it's reassuring to know that with all her programmed instinctive patterns the bee remains flexible, even wily enough, to outmaneuver her keepers.

The fresh pollen which you collect with a pollen trap can best be kept frozen in small bags in your freezer. To feed it in spring it is best mixed with a small amount of honey to give it consistency, pressed into thin cakes, and laid on top of the cluster. You can also buy natural pollen, certified to be uncontaminated by insecticides and to be from disease-free colonies. It is available through bee supply catalogues, and is also advertised in the bee journals.

Currently it sells for about one dollar per pound (so you can see the advantage of collecting and storing your own).

Much more economical and perhaps more practical for the amateur or beginner is to buy ready-made pollen substitutes. Their content varies, but generally they contain soy flour, brewer's yeast, and perhaps small amounts of natural pollen to increase the palatability. Often the pollen substitutes are provided with sulfa as a preventative medicine. Sometimes a commercial blend is mixed with special sugars and sold under a trade name. Such preparations are usually fed dry, or come already formed into cakes for direct introduction into the hives. Pollen substitute is sometimes provided at feeding stations in the middle of the bee yard where it is collected on demand, but of course this means the bees will not get it until the weather is warm enough that they can fly rather freely and regularly.

The particular brand which you buy will have instructions for best preparation of that compound and recommendations for feeding. Typically it is a dry powder, about the consistency of flour. And typically it is prepared by mixing it with sugar syrup or honey. Honey is best, both because of the bees' preference and because it retains its consistency better. Roll it out in a thin layer on a sheet of paper and place it inside the hive cover, above the cluster. Do not do as I did the first time I used it: I used a six-by-fourteen-inch sheet of cardboard, and spread on a layer of paste about half an inch thick. Obviously the hive cover would not close tightly on this glob of substance, and I had to do it over. Simply use a thin sheet of paper, and pat the pollen substitute out into a layer certainly of no more than one fourth of an inch. That will last the bees some time, and it takes no more than a minute to add another sheet when that one is used up.

When to begin feeding pollen substitute depends entirely on your geography and climate. Some southern beekeepers feed in December to build their colonies up for pollinization work in February. Here in Washington State, which has genuine winters but not so severe as those in the upper Midwest or New England states, I begin feeding in mid- to late February, depending on the weather. Counting twenty-one days for hatching of the first eggs laid, another twelve or fifteen days for maturing the new hatch, that means there will be young field bees ready by mid- or late March, in time to make a real haul from the first natural pollen. You will thus have a fine build-up by mid-April when the fruit blossoms come off.

Even if you have left your hives with ample winter stores, most likely you should also feed a quantity of sugar-water in the early spring, which for the hobby beekeeper is easiest done with a Boardman Feeder. Particularly if you are feeding pollen substitute to stimulate brood rearing you may find the colony to outrun its stores much more quickly than you anticipated simply because there are more bees to be fed. Even if the bees may not actually suffer want, if the stores begin to fall low before natural food is available, the queen may curtail her egg laying. Very often a hive which has otherwise wintered quite well will find itself caught short when a stretch of bad weather in early spring prevents foraging, with a flock of hungry bees and nothing to feed them. It is therefore very important to pay close attention to the condition of your hives during the crucial spring months. Hives may actually starve out during prolonged bad spring weather, but more likely you will simply lose the advantage of the spring build-up.

Feeding your bees syrup is little trouble and small expense, and always seems worthwhile. The syrup also provides a good opportunity to provide preventative medication—sulfathiazole or Terramycin. Again, these medications may be purchased through bee supply catalogues, and they come in different forms. Most are dissolved in syrup, although some are administered in dry powdered sugar. Some beekeepers advocate "out-feeding" syrup with medication so that wild bees (or untreated neighbor's bees) can be treated and thus reduce chance of infection reaching your own hives. Some beekeepers advocate "out-feeding" as a general practice, medicated or not, simply on the argument of convenience. The procedure in "out-feeding" is to set up a container (a plastic baby bath, for example; larger for a more extensive apiary) of syrup mixture, with plenty of slices of cork or chips of wood for the bees to rest on while tanking up. If you out-feed, however, place the containers at least one hundred yards from the nearest hives to prevent the feed serving to stimulate robbing.

Dry sugar may be fed, either plain sugar or specially prepared commercial varieties. It may be poured into the back of the hive in quantity, or spread along the top of the frames over the cluster. Dry sugar is not as effective in stimulating the rearing of brood as syrup and does not provide needed moisture. Sugar may also be made into a kind of rock candy and placed on the top of the frames over the cluster. This has the advantage of convenience, but may have the disadvantage of collecting moisture on the top of the hive which has been known to destroy a colony. Feeding candy is probably not advisable in very humid climates.

Sugar syrup is usually prepared by mixing dry sugar with hot water on the ratio of two parts sugar to one part water. Add a tiny bit of cream of tartar to retard crystallization. Sugar syrup can be made up as needed, or a quantity may be prepared at one time and stored.

The most common way to administer syrup other than with a Boardman Feeder is by using a tin can with a friction lid, and having a capacity of about half a gallon. Make a small circle of very small holes in the bottom with a nail or awl. When the syrup is added and the lid clamped tight, hold the can upright a short time until enough syrup drips out the holes to provide a vacuum. Subsequently the bees may feed through the perforations, but the syrup will not drip. To use, place an empty super on top of your hive and put the can of syrup inside, over the main cluster of bees. You will of course refill the can as needed. For winter feeding the empty space not occupied by the feeder may be filled with crumpled newspaper, etc. to conserve heat.

For a small number of hives I find the most convenient way to feed syrup is with the Boardman Feeder. The feeder consists of a wooden block shaped to be inserted into the entrance of the hive, hollowed to fit a perforated zinc lid, and with a channel providing a bee-way into the hive giving access to the perforations. The zinc lid fits a standard mason jar, usually a quart or half-gallon. When the Boardman

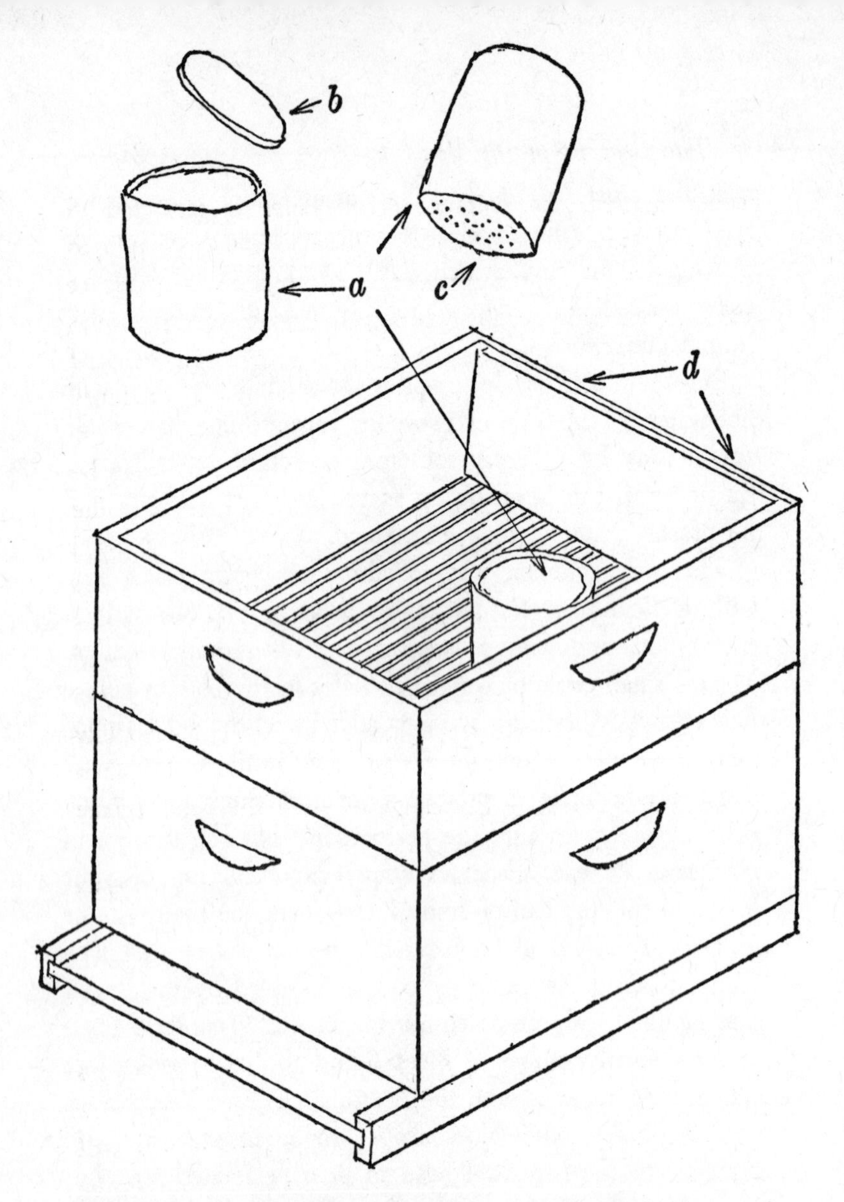

Feeding sugar syrup from a perforated can placed over the frames of the brood chamber protected by an empty, shallow super a. *tin can with tight-fitting friction lid* b. *Lid* c. *Bottom, perforated with small nail holes* d. *Empty super placed on hive body.*

Feeder is used, the entrance of the hive is restricted to allow only the smallest ingress. While it is very simple to feed with the Boardman Feeder, it has the disadvantage of attracting ants and possibly of encouraging robbing. A similar outside feeder is sometimes used which feeds from a special hole in the top of the hive. Robbing is supposedly reduced by this device.

There are also varieties of patent devices which may be used inside the hive. Generally they consist of containers the size and shape of a drawn comb, and a frame is removed to make room to fit them in place. This device also reduces the problem of robbing, but the advantage of such devices over home-made tin-can feeders is questionable. I would recommend that a beginning beekeeper start out using a Boardman Feeder. It is very simple and very handy. Only when your apiary consists of a number of hives, or when there are large apiaries located nearby, will robbing become a serious problem in the early spring or late fall.

How much to feed finally must depend on your own experience in your particular locale. The safest guide to begin with is simply to know that you cannot overfeed. Generally speaking the bees will take all you offer them, and assuming you quit when the season is sufficiently advanced that the bees obviously can support themselves, no difficulties should arise from feeding in quantity.

While the above instructions may sound as if you will spend half the time in the hives preparing for the spring build-up, actually you should not open the hives or leave them open more often than necessary during the late winter, and limit your intrusions to the milder days. Do not think,

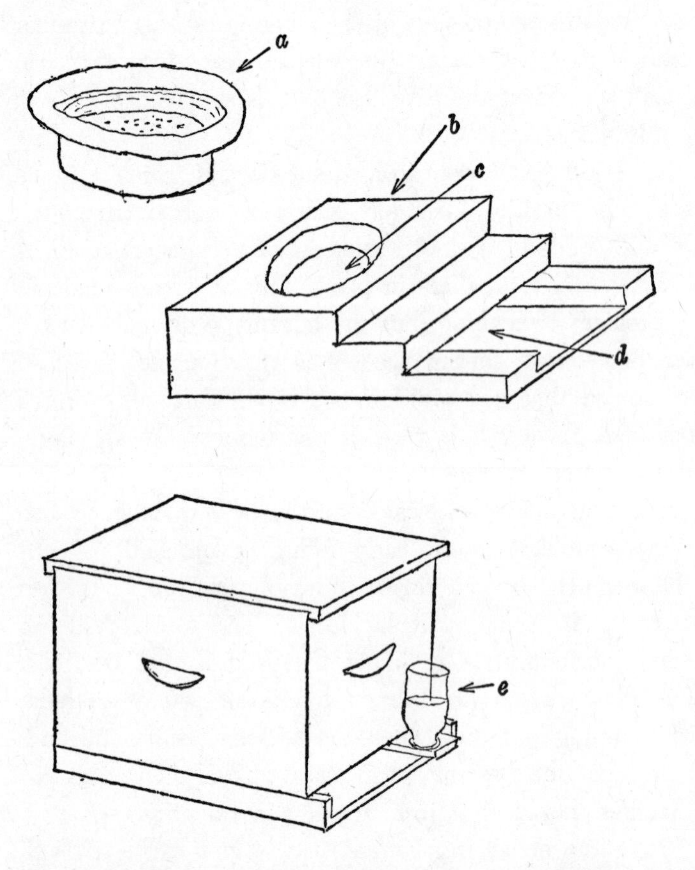

Use of Boardman Feeder a. *Perforated zinc cap, which screws on a quart jar* b. *Wooden portion of Boardman Feeder* c. *Opening into which* (a) *is fitted* d. *Channel (placed inside the hive) through which the bees gain access to the sugar syrup* e. *Boardman Feeder in place in front of the hive.*

by the way, that simply because it is less than mild weather the bees will be lethargic. You may be surprised to find that a hive you may have worked comfortably without veil or gloves during the last summer will fall on you like the wraiths of retribution when you open it on a mild day in late winter. For one thing, remember that now everybody is home; also conditions are just generally suitable to bring out the most peppery and protective side of their natures and dispositions.

In late winter, depending on locality, bees will begin to take "cleansing flights" when the temperature outside passes about 52 degrees. If the hives are situated with the entrance facing south the bees will be encouraged to take earliest possible advantage of fine days. If the entrance is shaded, their first flights may be delayed by days, even weeks, beyond what they would be with a more favorable situation. Cleansing flights are very important for the hygiene of the bees and of their hive. Bees much prefer to eliminate their feces on the wing, and may actually die rather than soil the hive. Occasionally unforeseen circumstances within the hive —fermentation of poorly cured honey, for example—may cause dysentery, which can be fatal to the entire colony if they are not able to take their cleansing flight. But the general advantages to the bees of these early flights need hardly be demonstrated. The bees also use these periods of mild weather, when they are able to fly outside, to begin cleaning the hive—removing dead bees, particles of wax, bits of debris, and the general domestic clutter produced by several months of enclosed living. And of course these outings also constitute orientation or familiarization flights for young bees hatched too late to have been out and around the previous fall before the weather closed in. You will notice that these bees will be predominantly young.

With increasingly mild weather the bees' activities will increase until, weather permitting, there seems almost to be normal activity. During this time the bees gather a great deal of water for domestic consumption, and pick around with the greatest ingenuity for early pollen sources. In many northern locations the first important pollen source is pussy willow, which is most productive of pollen. Other varieties of trees with most insignificant-appearing blossoms are heavy pollen producers; you'll find that you've never really been a good observer of flowers until you have begun to keep bees. Another very important and quite early source of pollen in most locales is dandelion. While the various early blossoms produce some nectar, the first really appreciable nectar flow comes from fruit blossoms. Ideally you should have managed your colonies by early feeding of syrup and pollen or pollen substitute so that when the fruit blossoms come off the colonies are at near peak strength. Commercial beekeepers who depend for some of their income on renting their bees for pollinization absolutely must have their colonies up to strength at that time. A usual part of the contract between fruit grower and beekeeper makes the strength of the colonies a point of stipulation in the agreement.

Do not look for a harvest of honey from the fruit blossoms, unless you are quite southern in location. This early nectar will for the most part be used up by the colony in their day-to-day economy and further build-up.

An unexpected danger period in the life of the colony

occurs regularly in many locales, irregularly in most, shortly after the fruit blossoms are harvested. There may seem to be lots of flowers, the weather may be very mild, and yet as far as the bees are concerned it may be a period of actual dearth. All flowers are by no means equal or uniform in their nectar yield, and weather conditions drastically influence the yield of the flowers. While there is little likelihood of a hive actually starving after the fruit blossoms have gone, a short period of slender pickings may cause the hive to reassess their condition and result in the queen slowing down or stopping her egg laying. Even a period of a few days' interruption can drastically reduce the field force available when the harvest season begins. Therefore you should keep track of the bees' stores during this period, and if necessary feed them again for a short period in mid-spring. This is most likely to be necessary or desirable if there is an extended period of rainy, cloudy, or windy weather.

What all beekeepers look forward to in the season of their hives is the main "honey flow"—a phrase with a piquant Old Testament flavor. Only personal experience will instruct you when to expect it and from what varieties of flowers it will come. A gardener may pride himself on the resplendent pasture his roses and lilies provide the bees. Actually a swarm would starve if it had to depend on the domestic forage of even a rather ambitious garden. It is not the lilies and roses and gladiolas that feed up a hive, although the bees do, of course, visit all of these blossoms. It is more likely to be the weeds the tidy gardener is at such pains to eradicate which provide the bees' bread and butter. Bees will find more sustenance along an untended ditch or roadway or fence row than in the apparently effulgent feast of a large flower garden. Honey plants constitute a complicated subject, one worth independent study and research.

In most locales the main honey flow begins in June, and with the care detailed above the colonies should be ready for it. The honey flow may come on all at once, or develop slowly over an extended period. Bad weather may shorten it or reduce its yield. Good conditions may provide a bonanza. The amount of honey a strong colony can process during the relatively short period of a good run is absolutely astonishing.

While the variety and quantity of blossom is of primary importance in the condition of the honey flow, weather conditions are only slightly less important. If weather is bad—chilly, rainy, or windy—the bees may be unable or unwilling to fly. Or in cloudy or windy weather it may be so late in the day when it warms up enough for the bees to begin flying that there is time to accomplish little. The production of the nectar itself by the blossoms is vitally influenced by the conditions of the weather. Influence is not always uniform or necessarily predictable, but generally speaking the ideal weather for an ample nectar flow is sunny, quite warm, and somewhat humid. Hot dry weather may be as unproductive as cold weather. Sometimes a light wind will reduce the productivity of an otherwise promising day. Flowers do not simply open up with their little drop of nectar in place and ready to induce bees to come along and help with the business of fertilization. A large number of conditions influence when (or even if) the nectar will be produced, and how much. Flowers require certain conditions for optimum fertility, and their production of nectar is geared and timed to coincide with these conditions. Flowers, after all, are in the seed business, not the nectar business. One variety of flower may begin yielding nectar fairly early in the morning; another variety in the same place on the same day may not produce until later in the afternoon. Bees learn to do their marketing when the shops are open, and may work one variety of blossom in the morning, a second in the middle of the day, and yet another in the afternoon.

Bees do have their preferences in blossoms, and individual bees even seem to learn to specialize. A given bee who has had her first success with a specific type of blossom may stick with that blossom during her entire career, if the flower's blooming season extends that long. For these reasons orchardists take pains to keep down weeds, particularly clovers, that may blossom in their orchards at the same time as the trees. A human being might infinitely prefer the great puffs of blossoms on the trees, but what appeals to a person is not at all necessarily what appeals to a bee. Bees are downright reluctant to work some blossoms, and farmers are especially diligent in removing plants from the area which might be more attractive. The cucumber is one of these species, and although bee action is crucial to their fertility, high yield may not always be assured even when large numbers of bees are in the area.

The yield from the honey flow also depends of course on the number of bees foraging a given area. If there are large numbers of bees in ratio to the number of blossoms, the production will clearly be reduced. A recent article in the *American Bee Journal* attributed the consistently low per-colony yield of honey in Poland to the fact that for virtually the entire country there are more bees than blossoms. Heavily farmed areas (excepting concentration on certain varieties of crops, such as legumes), tend to be able to support fewer colonies than less heavily farmed areas because the wild honey producing plants—the "weeds"—are not allowed to grow. This is the condition of the entire nation of Poland, and is true as well in large parts of this country. Bees have their own system of controlling the number of workers foraging in any given area. Studies have been made which show that the density of concentration of bees in a specific area never goes beyond what that area will support. If a bee arrives at an area already amply serviced by other bees, she will by the code of her craft be obliged to move on until she finds an unoccupied, or under-occupied area. This is why you never see bees wrangling or hassling over the nectar rights to a particular blossom.

The length of the honey flow of course varies by region. In the South the period may be quite long, but usually not particularly large, and the net production never particularly large, perhaps only slightly above the daily operating needs of the hive during the hottest months. Thus while the South has very long working seasons, a given colony tends to produce less surplus than a northern counterpart who has a very short but very rich season. So in some northern regions the main honey flow may run from mid-June, and during this short season most of the harvestable surplus is laid by. Most regions, however, have also a second, late, honey flow, derived from the rush of blooming that occurs from mid-August until frost. This can be a strong flow, perhaps even equal to the summer flow, but the honey tends usually to be darker and stronger flavored. Commercially it constitutes a lower grade and commands a lesser price than the early honey. Most beekeepers plan on using all or part of this late honey flow to build up the surplus the hives are to keep and use to winter on.

Properly managed bees should come through the fall with stores adequate for their winter needs. The additional special attention and care preparatory to winter is the source of endless arguments and productive of countless theories of systems among agriculturalists. In the past, beekeepers in areas of very severe winters "cellared" their bees; that is, they physically removed the hives from their stands and placed them in a cellar—properly stopped up to prevent their escaping from the hives, of course. The crucial conditions of cellaring bees are temperature and ventilation. Hives must be sufficiently closed to contain the bees, but at the same time the circulation of air must not be cut off. The bees must not be too warm, or they will be encouraged to be overactive. I've never lived in a region where cellaring bees was ever practiced, and my impression is that today it is rarely done. I believe that the development of modern practices of hive management has made it quite obsolete in this country.

Many beekeepers in areas of severe weather practice packing their hives for the winter. Various systems or techniques are used. Some wrap their hives with tar paper, some use tar paper as a loose wrapper, with leaves or straw being inserted between the wrapping and the hive. Some make a "false hive," a wooden box larger than the hive itself, which telescopes over it. Again various packing materials, leaves, etc., are inserted between the false hive and the hive itself to provide insulation. Various patent packing devices and cases are commercially available, made of cardboard and fiberglass, which slip over the hive for winter protection. These devices are reusable and with care can last many years. Sadly, some northern beekeepers find it most expedient and economical simply to kill their bees in the fall and restock their hives in the spring with package bees.

If you are located in an area of severe winters, the best advice is to check with your local agricultural extension service or contact the state bee inspector to find what local practices seem most successful. Except in those severe areas, however, little really needs be done other than to make sure that everything is shipshape, and that the hives have sufficient stores. A strong hive needs about sixty pounds of honey to go into winter. If it has less than that, surplus from a better provendered colony may be given. Otherwise the colony should be fed syrup. Often half-filled frames or poor-quality honey are given to a swarm with a questionable amount of provisions. Most likely, especially in a humid climate, a part of the winter preparation should be to provide a hole for ventilation in the second story of the hive. Commonly this is done by boring a hole into the front handgrip of the hive body. This hole allows circulation of air and prevents unhealthy build-up of moisture, especially if you expect the hive to be covered by snow.

Inevitably one's imagination or ingenuity suggests to him

using a heating cable or small light bulb to keep things cozy inside the hive during the coldest weather. The idea is good, but the practice can only be bad, unless outside weather conditions are themselves very mild. If the bees are kept warm, they will want to fly, and if they fly out when the air temperature is much below fifty degrees, they will simply die. If they were to fly out from a warmed hive to a temperature of zero, they would simply pelletize. Unseasonable flights can even be induced with a too-well-packed hive, according to reports. The colony itself generates a surprising amount of warmth; notice how quickly the roof of the hive, like the roof of a house, will melt away a patch of snow in the center. After all, the whole purpose of the honey stored inside the hive is to turn energy into heat.

When it comes, the snow actually provides in itself a healthy insulation around the hive. If possible, and most amateur beekeepers will find it possible, clear away the snow from the *front* of the hive only, down to the entrance. Clear away and open the entrance itself, especially taking care to remove any dead bees just inside the entryway, to provide ventilation. Leave the remainder of the snow around the hive for insulation.

In the cross fire of bee men writing and talking about wintering colonies we are frequently reminded that wild bees make it through the winter without packing, and few dead bees are found in late winter on the snow in front of a bee tree. The point is well taken, but also it must be taken into consideration that bees in unfavorable habitations *don't* survive; and further, that the colony has not had to share any of its honey with a beekeeper. The lesson is still basically a good one, though; make the living conditions of your colonies as good as possible, and make sure that they go into the winter with a good number of bees and sufficient stores. Very possibly packing or wrapping the hive will not prove really necessary.

7

Hive Management

Hive management seeks to achieve maximum strength per colony for the purpose of maximum yield per hive. Often an amateur beekeeper, however, actually seems to conceive of maximum hive strength as an end in itself. He takes private pleasure in the sight of a hive boiling with bees, and professional pride in showing it off to any beekeeping cronies he might have. While proper attention to wintering (assuming the hive has sufficient stores), and spring feeding constitute essential features to maintaining colony strength, a variety of additional techniques, wrinkles, and attentions to the needs of the hive are also important. Such attentions may not always be crucial to survival but may mean the difference between a mediocre performance and a superior one.

Perhaps the most important element of hive management has to do with swarm control. There is still a great deal to know about why bees swarm—which is another way of saying no one really has any idea why bees swarm. It is, of course, the natural way that colonies are increased. In their natural state probably a colony would swarm at least once a year, the swarm seeking and settling in a new location and becoming a separate entity, almost an organism. *How* a swarm develops is mostly clear enough; what specifically causes it to occur in a specific circumstance is still in many respects a mystery.

In the first place a hive "decides" to swarm. During certain seasons it is natural for swarming to take place—late May, all of June, even (despite the nursery rhyme) July. Obviously if it is to build up sufficient strength and store sufficient honey to survive the winter, the swarming has to take place early, before the main honey flow preferably.

Unless he wants to increase his colonies by the sometimes chancy means of collecting a swarm, the beekeeper usually seeks to prevent swarming. Perhaps no element of beekeeping arouses more commentary or wider divergence of points of view and opinion, or inspires more ingenious gimmickry, than techniques of swarm control. An obvious technique is to prevent the rearing of new queens—if new queens are not hatched, theoretically a swarm cannot possibly be cast. Unfortunately, this system has its drawbacks and is less than infallible anyway. To be even theoretically effective it requires at least bi-weekly scrutiny of each frame in the brood chamber—a not impossible task for the amateur, but a little unrealistic for a man keeping several hundred colonies. Further, the disruption of the hive activity cannot but be detrimental to production—both of honey and brood. These considerations aside, it is devilishly easy to overlook a queen cell, and for all your attentions have a new queen reared right under your nose. Many times after removing what I thought were all of the queen cells in a hive I have returned casually a day or so later to discover a nice ripe queen cell, situated obviously and prominently, which I had overlooked in my inspection. Last of all, even assuming diligence and thoroughness in removing the queen cells, the colony may swarm anyway; it is the old queen who leaves with the swarm, and the swarm may take wing and leave the colony queenless, having to make a new queen from scratch, so to speak, with eggs the old queen has left. What this means finally is, that if the colony is determined to swarm, it is going to swarm. The real science of swarm control seeks to anticipate and prevent this swarming impulse.

While the exact reasons for a hive swarming are obscure, other than the natural biological need for reproduction, there are clearly certain conditions which are conducive to swarming. The first of these conditions is crowding. In the natural state the room in the nest is finitely limited. When the hollow tree is filled with bees and comb, there just isn't any more space. The only logical course is to reduce crowding by the departure of a swarm. A second condition encouraging swarming is heat and lack of ventilation. Sometimes you will notice on a hot and sultry day large numbers of bees idle on the front of the hive, almost as if they were preparing to swarm. This is an alarming appearance, and means the hive deserves your immediate attention. If staying in the hive becomes unbearable because of the heat, it is most logical for the hive to look for healthier surroundings. Less frequently disease, the invasion of ants, or infestation of wax moths may be influential in stimulating a colony to swarm.

Most every beekeeper has his own pet theory about swarming and system for swarm prevention, and beekeepers pride themselves on the stability of their colonies as they do on the virtue of their sisters. The cold facts are that certain elements of the biological system are beyond the scope of managerial control. In even the best-managed apiaries, some swarming does occur, demur who will, sweeping the untidy facts under the rug.

The most reliable, reasonable, and practical systems of swarm control no doubt are those based essentially on space and ventilation, practiced with efficient queen management. First, give the hive an ample brood space. Two full hive bodies for brood rearing are not too much. Perhaps the first year you install your package bees in the hive you will only use a single brood chamber, but thereafter you should probably maintain two. If you are in a good spring build-up area, it is possible you may work up to two brood chambers the first year.

If the first requirement for swarm control is ample space for brood rearing, the second crucial need is for ample space for processing and storing surplus. Obviously this is what the whole business is all about—storing surplus honey—but a little undertanding is required to gauge when and how much storage room should be provided. The simple and obvious answer, just pile up a stack of supers at the beginning of the season, doesn't happen to be the right one. Two many supers given at one time will encourage the bees to follow a "chimney" pattern of filling the frames, using the center frames only. You also will be left with too many unfinished or partially finished frames of honey. It also seems to be disconcerting to the morale of the hive to have a vast, echoing empty warehouse above the tight compact cluster below in the brood chamber, and may in fact actually encourage rather than retard swarming. It is also an invitation to invasion by wax moths to have so much wax idle and untended by the bees.

The beekeeper's rule of thumb in supering is to keep one super ahead of the bees. That is, if they are filling one super, have another super already in place above them filled with foundation or drawn comb. It is important to the welfare of the colony generally as well as the specific psychology of having space available for future work. The ventilation of the hive is a complex process, and it requires adequate space for optimum circulation of the air. Space is also vital in the processing of the honey from the raw nectar. As it is brought from the field, nectar, as has been described elsewhere, consists of about eighty per cent moisture. The moisture is extracted largely by the process of evaporation. To remove the quantity of moisture necessary to process nectar into refined honey requires the availability of a rather astonishing quantity of air.

Further, the bees do not collect and store the honey cell by cell, frame by frame. They actually work simultaneously over extended surfaces of comb at once. At least during a heavy honey flow they do, and that is the time when swarm prevention is a premium consideration. The nectar which is brought in is not simply deposited in a cell and manipulated in that same cell until it constitutes honey. The field bee may drop it temporarily in a cell, and it will be subsequently retrieved, replaced, recollected, etc., many times before it comes to the end of the refining cycle. All this manipulation requires comb and surface considerably larger than the area required ultimately to store the quantity of honey which the process produces.

Therefore when you open a hive and find a half-filled super, it may appear that there is still plenty of room, when actually conditions of crowding may be becoming critical. Air for evaporation is lacking, room for merely handling the incoming product is crowded, and the frugal bee sees no additional space to generate optimism for the future. So the swarming instinct may be triggered, the cycle leading to the departure of the swarm irreversibly put into motion.

In addition to the specific requirements of ventilation for processing honey, the hive has more obvious general requirements for air. One of the most familiar summer sights in front of a hive is the arrangement of bees fanning with their wings to induce a flow of outside air into the hive. A little study of the arrangement of the ventilating bees makes it clear that the placement of the ranks assures that even on a still day the freshest (least moisture-laden as well as coolest) air will be collected and channeled inside the hive. Some of the bees will be positioned an inch or two from the entrance, fanning and moving the air, which is picked up by these nearer, and finally to those just inside the entrance itself. What you do not see from the outside is the arrangement of bees inside, completing the conduit which conducts and deploys this fresh and drier air through the interior of the hive, replacing that air which has been laden with the moisture from evaporating nectar. It is an astonishing system and arrangement, and clearly it takes a lot of bees to make it work. So again, if conditions reach a critical point and ventilation becomes more of a problem than the bees can handle, if ever bees may be thought of as "quitting," they quit, and collect in front of the hive for fresh air and relief from the super-humid interior.

In severe climate areas what can be done by the beekeeper to improve conditions of poor ventilation may be limited. When the temperature and humidity are both 100, manipulations of any sort will probably be futile. However, as soon as the colonies become permanently active in the spring (by the time of the fruit blossoms in the northern areas), all obstructions from the entrance to the hive should be removed. By the first of June at the latest you should reverse the bottom board of the hive to give them the maximum width of entrance. Keeping adequate supers of course helps the ventilation, but if you see evidence of poor ventilation indicated by clustering on the front of the hive, the opening may be made more spacious by tilting the hive backward on the bottom board and holding it in position by wedges. The cover may also be lifted to provide additional circulation of air.

There are also various manipulative techniques for swarm control mostly too complex for amateur (or professional!) employment, and some of dubious value. If you maintain two hive body brood chambers, reversing the order of the chambers is recommended by some. In the early spring, the queen generally goes up to the top of the hive, gradually working her way down as the season progresses. Reversing the order of the chambers puts the queen back on the bottom, with frames of brood above, and seems to confuse the swarming instinct. A variant on this principle, requiring more extensive manipulation, is called the "Demaree" system. This technique requires splitting up the frames of the two hive bodies, selecting those frames filled with brood and putting them in the second story, and putting the queen with several frames of brood directly below. In effect this splitting puts the hives into two sections directly above each other filled with brood, with two sections of empty combs (or foundation) in the other half of the two hive bodies. Presumably the queen will finish filling the combs of the bottom, move up and by that time the brood in the other half of the top section will have hatched, etc. Splitting the hives, as described elsewhere as a method of increasing your colonies, is a practice recommended by many for the sake of swarm control.

Probably there is no single answer to swarm control, and probably systems requiring excessive manipulation are more trouble than they are worth. The best "system" is most likely consideration of the basic biological causes of swarming, and taking them into account in simple practical hive management. Keep your colonies strong and free from pests, give them lots of room, and as much as possible look after their ventilation. You will not prevent swarming by such means, but swarming should be substantially reduced.

While supering is a necessary part of swarm control, it is also the bread and butter of honey production, to mix metaphors. To some extent the techniques of supering depend on your kind of honey production—chunk, comb, or extracted. By far the majority of amateurs concentrate on chunk honey production, and there are few real problems of supering here. First, of course, follow the general rule of keeping a super ahead of the bees. The amateur will proceed differently from the professional here, largely because he has more time and attention (probably) to devote to the project.

Normally the bees will begin by filling the center frames of the super, drawing comb to keep ahead of the incoming nectar. The bees tend to conform to a cluster configuration as they work the frames; that is, they tend to fill a football-shaped area, working gradually into the foundation beyond the edges of this parabolic configuration. They may fill the center frames completely, edge to edge, and only draw and fill the center, or adjacent frames. Particularly this tendency will occur during a light or sporadic flow. To discourage this practice (which may leave you with imperfect or practically filled frames), when the center two or three frames are completely drawn, and reasonably filled (not necessarily capped), lift them and replace them with the frames on either side. When those frames are drawn and filled, they may be in turn leapfrogged and replaced with partially worked or unworked frames. How this manipulation is carried out depends much on the honey flow; if the bees are drawing combs and filling frames rapidly, it may not be necessary at all. But it is often a help in producing the most attractively filled and whitest frames of chunk honey.

At any rate, when the super is about half filled, and honey is coming steadily, it is time to super. If you are extracting and using drawn comb in your supers, that is what you will give them; or you will supply the bees with heavy foundation. For chunk honey production, you will use supers fitted with thin foundation. Whatever product you aim for, place this empty super between the top brood chamber and the super in which the bees are currently working. Bees do not immediately take to the strange or the unfamiliar, so it helps accustom them to this new appendage to their establishment by forcing them to travel through to their working areas. In a short time they will begin drawing comb in the center frame, and soon begin storing honey. If supering is done too soon—either in terms of honey flow or the state of the previous super—the bees might at this point devote their energies to working on the new super and leave many imperfectly worked and incompletely filled frames in the old super. If such an event should occur, wait until the center frames of the new super are filled and rotate them with the imperfect frames of the previous super. If there is an active honey flow, in this way the bees can be cajoled into drawing most of the frames into perfect combs, and only perfect, or nearly perfect, combs are suitable for chunk honey.

When the time comes the third super can be applied in the same way—putting it between the brood chamber and the half-filled second super. And it is possible to proceed in this way through the entire honey crop. If you are producing chunk honey, however, leaving the supers in place the entire season tends to produce discoloration of the cappings—called *travel stain.* The discoloration is simply a result of the bees' tracking up the place, as a housewife would say, and has no effect at all on the actual quality of the honey. The appearance, however, is not at all as attractive as the clear pristine white of newly capped comb, and is worth your effort to avoid (which the amateur has time to do).

Therefore, when you add the third super, it is time to take steps to remove the first. If you have been successful in enticing the bees to fill each frame, edge to edge, fit a bee escape in position on an inner cover, place between the top super and the next to the top super, and within a few hours all of the bees will be gone, and you can carry the super away, frames and all. If the frames have not all been filled uniformly, you may want to remove only those frames which have been satisfactorily filled. Remove them, brush away the bees, put the partially filled frames in the center, and fill out the rest of the super with frames and foundation. The super may now be used as an "empty" super, putting it back between the working super and the brood chamber. Removing the frames as the bees fill them, by the way, is the best technique for producing the most attractive comb honey. It does necessitate frequent visits to the hives, however, and unless the manipulations are handled deftly it can cut into the total productivity of the hives to enter them so often.

Perhaps a word about the use of smoke and the smoker is appropriate at this point. Next to the veil, the smoker is the most familiar and recognizable tool of beekeeping; it functions by inducing the bees to stop whatever they are doing and fill up on honey. It has been conjectured that this is a built-in protective mechanism whereby in their wild state bees could abscond from a natural nest if threatened by fire and be prepared with sufficient emergency stores to establish a new colony. Whatever the primordial cause, that is what bees do when they are smoked: they fill up on honey, and a bee full of honey is a docile bee. The smoke by itself seems to have a soporific effect. There are substances sometimes used in bee smokers that have an actual narcotizing effect. A small amount of ammonium nitrate mixed in the fuel will render the entire colony unconscious for a short period. It is also said that there is a lasting effect of using this substance which makes even the most ill-tempered swarm mild and quiet until a new crop of bees mature. I have never tried this, and do not recommend it; if for some reason you want to try it, for example in removing a difficult swarm from an awkward clustering place, remember that this smoke is very hot, and if you blow it directly on the bees from a short distance it is apt to kill many. Other substances, from report, also stupefy the bees completely. British beekeepers sometimes dry the immature fruits of the common puffball, slice it, and use it in the fuel of their smokers. Supposedly it renders the colony temporarily senseless. It is also said that the dried leaves of wormwood, or absinthe, will render a swarm unconscious. Traditionally bees are reputed to have a natural antipathy for wormwood, and sometimes authorities say that to stop up the entrance of a hive with green leaves of wormwood will stop attacks by robber bees. This last, if true, would be very useful as a stratagem to know, especially if you happen to have a cluster of wormwood growing handy to your apiary.

In spite of this excursion into the exotic, the advised fuels for the smoker are cloth and the like. At best smoke interferes with the routine of the hive, and should be used as sparingly as possible. The actual use of smoke to its most efficient advantage comes only with practice. The general procedure is to blow two or three puffs into the opening of the hive, wait a moment, separate each of the sections of hive and its super and blow two or three puffs, lift the lid briefly and blow in a couple of puffs, then wait fifteen or twenty seconds. When you remove the lid, most of the bees will be calmed. As you work among the frames, an occasional puff of smoke will retain them at an optimum level of quiescence. Too much smoke can be worse than none at all, and exactly reverse the effect from what you intended. If the bees seem to be getting out of hand at any time when you are working with them, the best course is to close up the hive and return later. It will be more comfortable for you, and be much less injurious to the bees.

Beekeepers say that on a day you visit a hive with smoke you will have no net honey gain. Even if this might be a slight exaggeration, the less often you smoke, and the less smoke you use, the better. If you time your supering and manipulation of frames in the supers properly you most likely will need to use no smoke at all when you gain confidence and experience (and work on a bright sunny afternoon, during a honey flow, when most of the bees are not at home anyway). When you take off the honey a frame or two at a time, you will be able to remove most of the bees with a soft brush, and then take the honey immediately inside, where the bees cannot get to it. Bee brushes seem invariably to be made of soft horse hair. While these brushes are very convenient, animal hair seems less agreeable to the bees than fibers from other sources. Many beekeepers never use these brushes, but always depend on a leafy branch or bunch of grass casually collected around the hive. These vegetable fibers seem much less irritating to the bees, and I recommend trying such expedients. To me the effects of using a green branch have always seemed much happier. Most hobbiests, however, cannot resist the impulse to owning such an efficient and necessary-looking tool as a bee brush.

If you remove an entire filled super, it is possible, very late in the afternoon or evening, to lean it against the hive in the expectation the bees will all leave and crawl back into the hive in the chill of the evening. This is a dubious practice, since it is possible it may trigger a round of robbing if the honey flow is weakening. Or there may still be more of the daylight left than you really have accounted for, and the bees will remove enough of the honey to damage or ruin a good number of the frames. If the evening is warm, they just may prefer not to leave, and will still be found, happily prowling the combs, at midnight.

Supers may also be freed of bees by using a mild fumigant spread on a board, commonly an unused hive bottom. If you use this technique (which can work very well) be sure to follow the instructions for the particular fumigant to the letter, both for the safety of the bees and to protect the purity of the honey. Do not experiment with recipes or instructions you find in beekeepers' manuals; contaminated honey is too dangerous, and live bees too precious, to afford risks. Commercial beekeepers often use a blower, gas- or electric-powered, which forces a stream of air through the super, rapidly clearing it of every last bee. These devices are moderately expensive, at least for the scale of operations the usual amateur beekeeper is engaged in. Probably the best system for removing bees from a filled super, finally, is the bee escape; it requires a modicum of patience, but amateur beekeepers must have patience or they would never have pursued the craft in the first place.

One of the most important margins between success and mediocrity in hive management is involved with queen management. A young, prolific queen is one of the best possible hedges against swarming. When you buy your packaged bees from a reliable professional, you are pretty sure of getting a good queen. Most likely you have ordered your bees and queens according to variety or strain, and the young queen will almost certainly be vigorous and prolific the first season. However, a queen is not always a queen; among professional beekeepers, queen judging is as exact a science as sire judging among horse breeders or cattlemen. The ideal queen is judged partly by appearance—it's nice to know that she should *look* like a queen—partly by judging her observable laying habits. The minuscule, banana-shaped eggs should be deposited square down in the bottom of the cell, adhering from one end, and on a perpendicular to the bottom of the cell. The queen should lay in a solid, elliptical pattern around the comb—"lay like a board," as the beekeepers say. Opinions vary about how many eggs a queen can lay in a day; estimates differ enormously. Some maintain she can lay up to six thousand eggs a day, others say her maximum is nearer twenty-five hundred, and a common estimate is that a thousand eggs a day is a good practical figure. You won't be counting them, anyway. The important thing to observe is that the comb of brood ends up solid (calculate about twenty-eight cells per square inch, if you're interested in figures). A frame which is only partially filled, or one in which the brood is patchy and irregular, likely indicates that the queen may be failing, and should be replaced.

Most likely your queen will perform admirably her first year, and most likely her second year as well. Her active life may extend as long as five or six years, but she should never be kept that long. If her reproduction falls off seriously, generally she will be superseded by the decision of the colony, or however such decisions are made, by the simple expedient of making a new queen and disposing of the old. If the hive swarms, the old queen will leave with the swarm, and the colony will have automatically requeened itself with a young and vigorous queen from its own ranks.

There are certain indications by which even an amateur beekeeper can tell he has a failing queen. No brood at all, of course, is the most obvious and most traumatizing symptom. An excess of drone brood may indicate a failing queen—one who has lost her fertility and can lay only male eggs—or a sterile or failing queen being supplanted by laying workers. But the beekeeper should also respond to the more subtle symptoms indicated by brood frames only partially filled, frames with narrow bands of brood and large areas of empty cells, or random patterns of eggs being laid. Close examination may reveal eggs not normally (i.e., perpendicular, at the bottom) positioned in the cells, or even more than one egg per cell. Such particulars all suggest that the queen's procreative powers are declining.

The easiest and surest technique of assuring a continuing productive queen is periodic requeening. Theories vary as to the schedule; some beekeepers favor requeening every three years unless circumstances indicate a given queen should be replaced earlier. Some believe requeening should be done on alternate years, and some maintain queens should be replaced annually. To keep track of their queens, beekeepers often order them marked, as recommended here elsewhere—sometimes marking is done with an actual spot of color, sometimes by clipping one wing. Many beekeepers find it useful to maintain a record of each hive, such as the queen's vital statistics, on the inside of the hive cover. This is quick and efficient, and there is much to recommend it.

Considering what is at stake—the whole progeny of the hive—the principle of annual requeening is certainly worth considering. It is a bother, to be sure, and an expense if very many hives are involved, but the results can hardly ever be other than advantageous. The question arises, then, respecting the optimum time for requeening, whatever schedule is followed. Routine requeening is best done in the fall or spring; requeening at these times entails minimum disruption, and the likelihood of failure is small. If you requeen in the spring, usually you can see immediately whether the operation has been successful, so to speak, by the almost immediate appearance of eggs, and then brood. If you requeen in the fall, you have the assurance of the queen being in place and laying at the earliest possible time, being at her maximum fecundity at precisely the most crucial time in the colony's build-up. Further, if spring inspection fails to show brood and eggs, there is ample time to get a replacement without a serious setback. All things considered, I prefer the fall for scheduled requeening, although there are many options to the contrary; most likely the difference in advantage of one over the other is less than crucial. Signs of a failing queen, however, should be treated immediately regardless of when it occurs during the year.

There is always the alternative of letting the bees take care of the matter of a queen themselves. Sometimes, however, even with this plan there are times when the bees cannot or will not take care of themselves, and the problem may proceed into the critical stage before it is detected. The most drastic condition occurs when you have a dead queen and drone-laying workers. Detection of this condition usually means that matters are already crucial. Since most workers are capable of parthenogenic reproduction, when a queen fails or dies certain workers take over her egg-laying duties; in an advanced condition, large portions of the brood chamber may be made over and the cells reconstructed to the larger size required by drone brood. The drone cannot, of course, work, and it is merely a matter of time until such a colony, *drone layers,* as they are called, ceases to exist. Pathetically in such a condition the workers may still construct queen cells, laying drone eggs in them in the attempt against all odds to produce a queen.

To requeen a drone-laying colony is not easy, however, and any techniques are less than foolproof. It is said that more such colonies are lost than are ever saved. Some beekeepers recommend not even attempting it; rather, to destroy the existing bees (and all comb made over into drone cells) and replace them with packaged bees and a new queen. What causes the problem of re-establishing a queen into a laying capacity—making the hive "queen-right," as the expression goes—is apparently that the worker or workers who take over the egg-laying practice become the "queens" of the colony. Perhaps their glands begin to secrete the substance which identifies the queen, and makes the colony accept them as heads of the colony. There is no way for the beekeeper to identify such a bee and remove it. Trying to introduce a new queen into such a colony is like trying to introduce a new queen into a colony which is already queen-right; it simply will not be accepted. Consequently various techniques of sneaking the new queen by, so to speak, have been developed, and practiced with varying reports of success. One such technique is to take all of the frames outside the hive and shake off all the bees some five or six feet in front of the hive, replace the frames and introduce the caged queen. Reportedly, when the bees return to the hive, the chances are good the new queen will be released and accepted with no further complications. Perhaps the egg-laying workers are physically incapacitated by their procreative activities, and thus cannot make their way back to the hive. Or perhaps the trauma alone causes a reshuffling of the social order, a precarious balance under all circumstances, which makes conditions favorable for the acceptance of the new matriarch. Or the laying workers may never have made orientation flights, and thus cannot find their way home. At any rate, the system reports good success.

Another technique recommends placing the queen in the

hive with certain remodeling of the queen cage. Cover the candy plug so that it is completely inaccessible. Remove the other cork (without, of course, allowing the queen to escape) and attach a slim, long wire to the cork. The wire will be conducted outside the hive, with sufficient length for a hand hold to allow a pull from outside the hive to remove the plug. The cage must be attached *firmly* to the frames, so pressure on the wire can remove the plug without dislodging the queen cage. The queen cage, thus modified, is allowed to remain untouched in the colony of drone layers for five days (they cannot, of course, get to her, since the candy plug has been made inaccessible). During this time the colony will have become accustomed to feeding her through the screen of her cage, she will have completely assumed the hive smell, and the workers will have become used to having an actual queen in the colony again. After five days, then, the wire is pulled from the outside, without opening or in any other way molesting the colony, the cork drawn, and the queen instantly released. Most likely she will begin laying within minutes. Using this system, probability of acceptance in a drone laying colony is reportedly very good.

Requeening an otherwise queen-right colony is not difficult. If possible remove the old queen about twenty-four hours before introducing the new caged queen. There is nothing that produces consternation in a hive equal to the absence of a queen. It is believed that minute traces of the "queen substance," the source of the queen's peculiar identifying odor, are constantly passed about among the bees in the hive through their incessant touching of one another. When no "queen substance" is present, the fact becomes known in a remarkably short time. Activity comes to an abrupt cessation, and all hands seem to devote themselves principally to lamentation, or its equivalent. If eggs or newly hatched larvae are available, they will shortly begin constructing a queen cell. Otherwise production will proceed at a reduced and lackluster rate. To expose the hive to twenty-four hours of queenlessness intensifies their relief at having a queen again when you introduce the caged queen, and makes her acceptance by the "governing board" of the colony virtually assured.

Aside from periodic, scheduled requeening, other non-seasonal occasions will sometimes indicate requeening is advisable. A surprising number of queens are inadvertently killed or materially injured unintentionally in the periodic manipulations of the hive. In midseason, therefore, your hive may suddenly be queenless, or worse, in possession of a marginal queen. Fortunately, queens may be ordered from the breeders well into the fall, shipped airmail, and can be in hand soon after an unfavorable condition is detected. From most suppliers you can order by phone, if you feel your urgency justifies the expense.

If you are maintaining a particular species or variety of bee, sooner or later you will encounter the problem of sustaining your strain, unless you turn into a breeder in your own right. Following exactly the Mendelian principles of selectivity—certain strains are dominant, certain ones recessive. Being parthenogenic complicates matters; the drone eggs a queen lays will all be of the type of the dominant genetic force in the queen's lineage. If you have a hybrid strain, midnight for example, her drone eggs will become Italian bees. Her queen progeny, if they mate with Italian sires (her own, or tramps), will then produce only Italian workers. If you buy a midnight queen who has mated with an Italian drone, her progeny will all be Italian. Practically, then, this means it is possible to keep a particular queen and find she does not throw true, that her progeny are not of her own kind.

Such accidents may be avoided by buying "tested queens," queens which have been kept by the queen breeders until it has been demonstrated that they lay fertile eggs which hatch into bees true to type. Tested queens cost at least double the price of untested queens, but are usually available, if you choose them.

Most likely the problems of maintaining strain comes in the hive itself, after a queen has been introduced and seemingly accepted. Much more often than realized a queen is not accepted completely when establishing a packaged colony. Perhaps the queen has been injured in transit, or proves unsuitable in fecundity, is damaged after she is released into the hive, or simply for reasons only known to the bees themselves is just not accepted as the matriarch of the hive. In such cases, a queen may lay a number of eggs, or even become fairly well established, but the workers nevertheless will immediately construct supersedure cells and make a new queen. Thus without the beekeeper even knowing anything is untoward, by the time the first brood of his newly established colony has hatched the colony will have quietly requeened itself and destroyed its unwanted original queen. Under such circumstances of the hive requeening itself, the chances are more than good that the strain of bees the beekeeper ends up with are not the same as he intended, unless he began with Italian. Only if the new queen mates with a drone from the hive will the progeny throw anything close to the original strain. Because they are by far the most common bees, and by far the most tramp drones are Italian, the chances are the hive will ultimately end up Italian.

Even if you buy a tested queen, then, the chances are less than perfect that your strain will remain true to species for even the first year. Immediate supersedure occurs distressingly frequently. It is about as likely that a tested queen will be superseded as an untested (although there seems to be some evidence that an experienced layer is more acceptable and therefore less liable to supersedure than an untested one). While for any amateur beekeeper the added expense may be comparatively small, it probably is not worthwhile, all things considered, to go to the bother of specifically ordering a tested queen. Unfortunately you may not detect the failure of your own desired bloodlines until June, and it is best not to requeen during the honey flow. Probably you should simply live with your queen regardless until autumn, and requeen then, back to your preferred species.

Requeening is also usually indicated as the sensible solution when you find yourself possessed of a bad-tempered swarm. You will quickly learn to predict the reactions of your several colonies (if you maintain more than one) to your periodic tamperings and molestations. Some will be docile and even companionable; some will sting you even if conditions are most favorable and you present yourself clad in armor. The key to the colony's disposition is the queen. Get rid of the queen and you will most likely get rid of the bad temper. If you requeen annually you may more than likely prefer to live with a disgruntled and disgruntling swarm than requeen in midseason. The productivity of a colony, its wintering qualities, and certain other considerations are also dependent to an appreciable measure upon the queen. Some commercial beekeepers who rear their own queens according to planned and controlled systems find the production of their strains increasing by nearly half over the course of several years of selective breeding.

If you capture a "wild" swarm, most likely you will want to requeen it as soon as possible, regardless of the season. When you capture a swarm, the major consideration will be to build it up to survive the winter, with no thought of obtaining any sort of surplus from it the first year. When you capture a swarm, therefore, the quicker you can give it a queen from known bloodlines, the better. By the same token, if you buy hives or take over swarms from houses (perish the thought!) by all means requeen as soon as possible.

If you maintain a consistent strain of bees in your own bee yard, it is more than likely that your virgin queens will mate with domestic drones. Therefore domestic, "natural," requeening is quite possible even without maintaining a careful or programmatic system of queen breeding. To requeen you may simply remove and destroy the existing queen—making sure of course, that the hive has eggs or freshly hatched larvae so a new queen may be made. Or in the case of a failing or drone-laying queen, a frame of brood containing a ripe queen cell may be introduced from another colony. Easiest of all—but not necessarily best—you can simply let nature take her course, and depend on the bees requeening themselves as they see fit. But casualness easily leads to laxness, and laxness leads to extinction. At best, bad or indifferent queen management leads to mediocrity; at worst, it will put you out of beekeeping entirely.

8

Honey Husbandry

THE MANIPULATIONS, and even the ends, of the amateur beekeeper frequently differ from those of the commercial beekeeper; indeed, may even be at odds with his. The amateur keeps bees principally for pleasure and diversion; he wants and demands a harvest of honey, but here again his interests are amateurish. He wants first of all quality—taste and appearance. He is pleased by quantity, but more as a measure of his competence than as an end in itself—similar to the marks your children bring home from school.

Even though commercial beekeepers seem more capable of maintaining the keen pleasure of amateurism in their occupation than most, they are after all in business. Many of the manipulations of an amateur would be unfeasible in a large-scale operation—or too chancy or even potentially detrimental to economical production. An amateur might split two or three hives primarily to see if he could do it; he would get honey adequate to his needs even if the operation didn't work well—or at all. The professional most likely would not care for the risks to his bees, the expenditures of time, the tie-up of equipment, or the risk of profit.

The commercial keeper most likely expects income from at least two sources from his colonies, fees for pollinization, and sales of honey. From a number of people I have heard expressions of surprise that the orchardist is expected to pay for the service of the bees, rather than the other way around. This reaction always reminds me of the spinster who wished to have her Siamese cat present her with a litter of kittens. Following up an advertisement for stud service, she said with astonishment, "You mean *Suki* has to pay?"

It has been repeatedly and dramatically demonstrated how dependent the production of fruit, etc., is upon visitation by bees, and most farmers who specialize in crops which depend on bees for pollinization are unwilling to take up the sideline of beekeeping on sufficient scale to do themselves any good. So the fee for the service of their bees is an essential part of many beekeeper's income—indeed, the primary source for some, who practice mobile beekeeping by following pollinization northward. As a footnote to this matter, in the last few years there has been experimentation with *disposable bees,* perhaps the ultimate in throw-away economy. "Disposable bees" are units of packaged bees provided in a weatherproofed cardboard box, with a queen, which are scattered about the area needing to be pollinized. A tab is removed from the front of the box, allowing the bees to fly free; when their usefulness is finished, they are gassed. I find it impossible to generate the slightest bit of enthusiasm for this practice.

For the beekeeper who rents his colonies to pollinize orchards, the actual harvest of honey from the blossoms is commercially negligible. It is generally consumed by the hive itself in continuing its build-up to the major honey flow, from which the real surplus is harvested. The amateur beekeeper who has been able to build his colonies up to nearly peak strength by fruit-blossom time may be able to set aside a little of this fruit-blossom honey, particularly if he is willing to provide additional feeding of syrup in compensation. Certain early blossoms have sufficient reputation for unique excellence that beekeepers find it profitable to manage their hive to take some of this honey off, again usually by providing supplemental feeding. The wild raspberry is one of these blossoms. Although I have never tasted it, it enjoys an epicurean fame. If you are planning to obtain fruit-blossom honey, you must bear in mind that administration of medication must finish a month before the harvest. In such a case medication most likely should be restricted to fall.

Professional beekeepers will regularly—or irregularly—plan on moving their hives more than once during the season to take advantage of the peak of different honey crops. Some managers regularly move their hives several hundred miles, with intermediate stops in between. It has become so popular for northern beekeepers to winter their colonies in Florida, and give themselves a sunny vacation at the same time, that that state is trying to enact legislation regulating or taxing touring bees. Beekeepers maintaining stocks of bees in commercial quantity generally locate their apiaries in dispersed locations, the number of hives at any given location depending entirely on what the traffic will bear, a matter chiefly dependent on the beekeepers' judgment and experience. Colonies so maintained are called "out-apiaries," and sometimes the beekeeper will rent (or buy) a small piece of land which he can fence and maintain to accommodate these out-apiaries.

Many surburban amateurs will no doubt find similar expediencies necessary. He may winter a few colonies in his back yard, perhaps even see them into the spring build-up, but the suburban flow may be inadequate or of too poor a quality, for an adequate or acceptable surplus. Or the suburban beekeeper may find his neighbors' wives attributing every winged insect she finds in her kitchen to his apiary. Except in the most disastrously urbanized areas of our country, someone who seriously wants to try his hand should be able to find a location within practical driving distance where he may make arrangements to establish a modest out-apiary of two or three hives, perhaps even on a year-round basis. It is not as much fun as having the bees right at your elbow, but it is a compromise preferable to being bee-less.

If you do find it necessary to shuttle your bees from your back yard to an out-apiary, moving the bees really is no major undertaking. The range of a bee, while extensive considering his size and vulnerability, is fairly short. His range will likely be no more than two and a half miles, and probably a great deal less. In fact, if a bee has to fly more than a quarter of a mile to reach his regular honey source, the surplus collected will be seriously reduced. The bee finds his way home the same way you or I do, by following recognizable landmarks until she arrives at the spot where the opening of her hive ought to be. Therefore when you move a hive some distance, the bees will have to learn an entire new system of landmarks, and thus will immediately program themselves to find their new home. If you do *not* move your hives beyond where the maximum flight radius of the old and the new location meet, bees stumble on old landmarks, become disoriented, and try to fly back to the original location. Of course all such bees would be irretrievably lost. Therefore if you plan to move your bees, you must arrange to move them at least five miles to be assured that no bees will be lost.

When you are moving the hives only a short distance (but of course more than five miles) preparations are relatively simple. Fasten the hive securely to its top and bottom either by nailing or using metal bands. After dark the night before the move close off the entrance of the hive—*securely!* If you can move them that same night, so much the better. If not, move them as early the following day as you can. The major consideration in moving the hives is ventilation. The bees can suffocate very quickly in hot weather. If it is going to be necessary to keep the hive completely closed for more than an hour or two after daylight, you should close off the opening with screen rather than a solid block in order to allow some circulation of air. If it is necessary to move them a long distance, arrangements should be made which allow screen to replace the top and bottom boards for ventilation. Most amateurs will most likely never be involved with such a move. Commercial beekeepers, moving bees long distance by the truck load, have to take elaborate measures, however, to ensure the safety of the bees. Some commercial beekeepers use what amounts virtually to permanent mounts on trailers, which can be parked at a new location, the hives opened, and the bees sent out to work without ever unloading them. In some southern locations bees are moved around rivers on barges, an extremely old practice. In ancient Egypt, as illustrated by stone carving and friezes, pottery hives were maneuvered down the Nile on boats following the honey flow as the season advanced. Perhaps the practice is still followed there.

When you are ready for the honey flow, or rather when your bees and the flowers are both in agreement that the time has come, you may want to install a queen excluder between the brood nest and the supers in which the surplus is to be stored. This is essentially a screening device, usually made from a sheet of stamped zinc, or else zinc-plated metal rods. The openings between the interstices are large enough to permit the free passage of the workers, but too small for the larger body of the queen. Its advantage is that it assures you of having no brood in the frames from which you hope to harvest honey. If you are producing comb honey, the rearing of even one cycle of brood in a comb effectively ruins it forever. The disadvantages of using a queen excluder is that it impedes the free passage of the bees; even though they can move back and forth through the openings, it is somewhat inconvenient and to that degree detracts from efficiency. Apparently the passage of the bees through the screen tends to add to the wear and tear of their wings, and is perhaps a factor in reducing their effec-

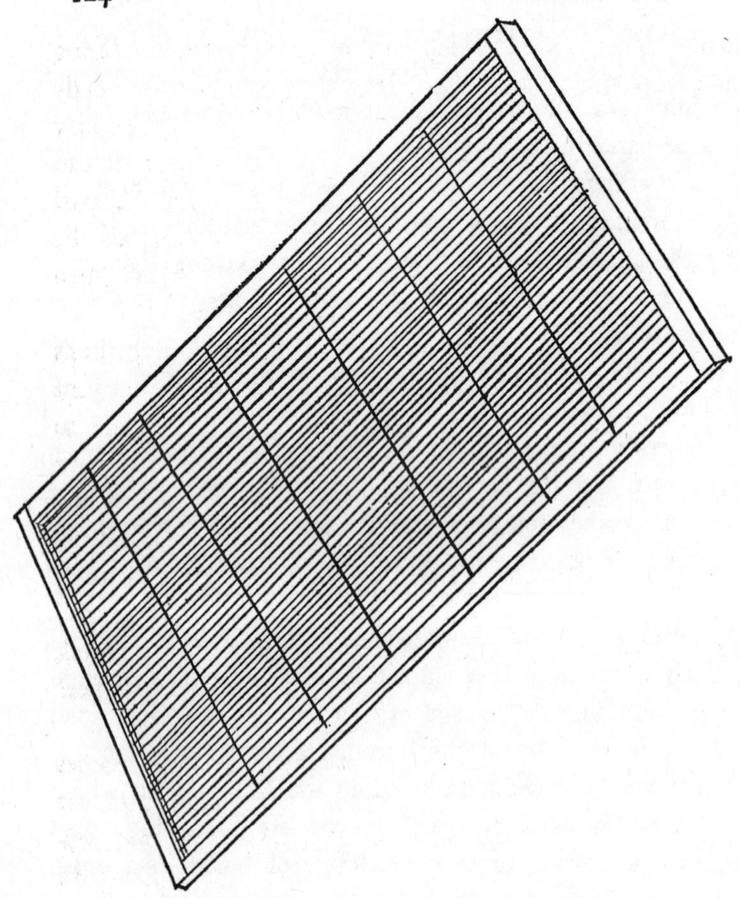

One kind of queen excluder—welded rod construction

tive field life. It's a tossup finally whether you lose more honey through the inefficiency the queen excluder imposes, or through loss because brood is reared in the frames. I'm inclined to believe the former, although many beekeepers do not agree, and I seldom use a queen excluder during the main part of the season. If you have a two-hive body brood chamber, well filled with drawn comb, the queen may never be tempted to go higher. If she does, let her. It can happen, of course, that the queen may just make a pass up the full height of your colony, laying eggs in all the center frames. In that case you should have used a queen excluder.

Much of the foregoing has been based on the hypothesis that your intentions are to produce chunk honey. If you want to produce comb honey in those little boxes which are so attractive, but so expensive you never buy them, you have different and slightly more difficult problems. Let me say that unless you have commercial aspirations there is little advantage to these small one-pound frames other than their rather cunning petiteness of appearance. In themselves they are no more convenient to keep or use than cut chunk honey, perhaps less, and they are less efficient for the bees to use, and therefore less productive.

Like the queen excluder, the nature of these box frames restricts movement within the super. It is necessary to use a special super arrangement, which sets up wooden separators between the box frames. The bees tend not to like to work in these boxes, and they are best given to the bees during a good honey flow, when they have already partially filled a regular super. The super containing the box frames is put between the hive body and the super being worked. You must expect that not all of these box frames will be perfectly filled, but experienced beekeepers can time supering and manipulate the supers to produce mostly perfect combs. Imperfect combs are consumed in private (I suspect often in secret), by the beekeeper and trusted members of his family.

When you take off the honey a few frames at a time it is possible to keep varieties of honey separate from each other. This is of little interest perhaps to most commercial beekeepers, but is of considerable interest to many amateurs. In point of fact most honey is mixed anyway. The multiple handling of the nectar during the reduction and curing process assures that if more than one variety of nectar is coming into the hive, the resultant honey is necessarily mixed. The habit of bees is to work one variety of flower at a time, and the fact that bees do have particular preferences does make it possible to say with reasonable precision that such and such a honey is, for example, *clover*, or *bass wood*, or *fireweed*, etc. There is little means of testing other than color or taste, what a particular variety of honey was made from. Microscopic examination may reveal pollen particles, which may in turn be identified as to origin, providing a good but still not totally infallible indicator of the nectar source. Pollen has a way of getting around in a hive, and it would not be surprising if apple pollen appeared in a frame of "clover" honey.

At any rate, many amateur beekeepers like to attempt keeping their honey varieties as distinct as possible, removing the frames as they are capped, and writing on the top of the frame both the date and the principle variety. Oh yes—how do you *know* what the nectar sources at the given times are? You watch the bees, of course.

What you do with your honey after removing it from the

hives can present serious problems if you have considerable quantity. Obviously it must be stored in a "bee-tight" location. Bees are incredibly ingenious at finding their way to honey, particularly during a slack period. Aside from the fact that they will steal you blind, the discovery of a hoard of untended honey almost always triggers a wave of robbing mania, which may result in serious harm to your apiary. You must also protect your honey from other insects and vermin—ants, flies, mice, earwigs, etc.

The honey may be stored either in the frames or cut from the frames and placed between sheets of waxed paper or plastic film. Or the combs may be cut and placed in jars, cans, plastic boxes, etc. I cut many sections and place them in square plastic boxes for Christmas presents. Plastic freezer boxes are ideal and quite inexpensive.

So far I have talked exclusively of producing comb honey. For the beekeeper with only a few hives this is much the most convenient and economical method of production, and most people will agree that comb honey is quite superior to strained honey. However, for an operation large enough to justify the expense, there are many advantages to producing liquid or strained honey by extracting it from the combs, and leaving the combs intact and empty in their frames. These extracted combs may be immediately replaced in the hives during a honey flow and will be speedily refilled at enormous saving of labor to the bees and efficiency in the economy of the hive. A great deal of honey must be consumed to make beeswax—estimates run as high as twenty pounds of honey to produce one pound of wax. A more conservative and likely figure is around six pounds of honey per pound of wax. The advantage to honey production for bees to have drawn comb available for immediate reuse is obvious. Curiously, such an authority on practical beekeeping as Mr. Latham argues that this savings is largely illusory, and while such a quantity of honey is indeed required to make wax, there is no converse savings if they are not required to build new comb. The bodies of bees prepare wax when the bee is of a certain maturity whether it is used or not. Mr. Latham opinions that it is better to crush and strain the combs and add the wax to the profits of the apiary. Few if any commercial beekeepers observe this practice, however.

If the combs are not replaced immediately on the hives to be refilled, it is wise to allow the bees to clean them before storing the supers away for any length of time. The usual practice is to stack a super or supers of extracted combs on a hive, between the hive body and second brood chamber. The bees will quickly clear the combs entirely. If the frames are stored away "wet," there is a serious risk they will mold and be rendered entirely useless. On no account should the supers be simply placed in the bee yard to be cleaned; this is an open invitation to a round of robbing.

After being cleaned by the bees, it is a good idea to fumigate the combs to prevent infestation from wax moths. Many commercial preparations are available, advertised in the bee journals or catalogues of the various bee supply houses. As in the case of any chemical preparation to be used around bees or their equipment, read carefully and follow religiously the instruction for the particular preparation you use. The supers of extracted comb should be stored in a bee-tight storage room, protected from pests and vermin. It is also possible to store them outside for the winter, in stacks protected from the ground, and tightly covered.

The wax in extracted combs quickly comes to take on a dark hue. The comb itself is actually improved by reusing, harder and less inclined to buckle or slip in the extractor. Comb in which brood is raised is even stronger, and some beekeepers make a point of seeing their extractor frames through one cycle of brood rearing. That brood has been reared in the comb in no way affects the quality of the honey or its flavor.

How you handle your honey once it has been extracted depends partly on quantity and partly on what you intend to do with it. As it comes from the extractor, the honey will have many wax particles suspended in it. These particles are completely wholesome, of course, but detract from the clear, sparkling appearance one expects in commercial products. If you want to remove these particles, it is necessary to use a settling tank, or similar device which allows the honey to be drawn off from the bottom after the wax has floated to the surface. The honey may be heated to about 140 degrees (no more) to speed up the process. While I'm not a food-faddist or health-food addict, I'm inclined to believe honey should not be heated at all, and would never trouble to clarify extracted honey. The clarity of honey from certain varieties of flowers is troubled by persistent, almost permanent, air bubbles. A notable variety of this sort little likely to trouble American honey producers is heather honey. This honey, especially popular with the

British, is so viscid that it may not be extracted by the centrifugal extractor, but is always obtained by crushing and draining the combs. Heather honey thus produced always has in suspension minuscule air bubbles, which have of course no effect upon the flavor, but impart a cloudiness to the honey.

Packaging extracted or strained honey is not likely to pose a very significant problem for the hobbiest. If you wish, you may buy through various bee supply outlets honey containers of varying sizes and shapes—glass, plastic, and tin. The same supply houses also furnish varieties of labels, if you must have them.

In time virtually all honey will crystallize. The honey from some sources—alfalfa, for instance—may crystallize in a matter of weeks. Others may crystallize after several months or longer. The size of the crystals and the consistency of the crystallized honey depends to some extent on the variety also. Some will crystallize hard, while some will take solid form in a cream-textured confection. A considerable quantity of honey is by preference marketed in crystallized form—called "spun honey," "honey butter," etc.

A peculiarity of honey's crystalline form is that it can largely be influenced by "seeding" liquid honey with crystals from a batch with a desirable consistency. If crystallized honey is added to liquid honey, it provides a catalyzing action, and crystallization proceeds very rapidly. It is exactly the same principle as the familiar experiment in high school physics where a super-saturated solution is induced to crystallize by seeding it with formed crystals. Or cloud seeding to produce rain, for that matter.

Some people greatly prefer this crystallized honey, sufficiently to pay premium prices for it. But if you have crystallized honey that you wish to reliquefy, this may be done easily by setting the honey container in hot (not boiling) water. It will reliquefy quickly. Do not get the honey too hot, however, or the flavor will be impaired.

Although you may have used the idea of realizing a sideline profit from honey as a rationalization for starting beekeeping, more than likely you as hobbiest will never sell any, or hardly any. Honey is such a pleasant thing to give to people—like flowers—that even one who intends to sell his surplus usually soon finds himself surplusless.

9

Enlarging Your Apiary

Beekeepers and their colonies are somewhat like bees and their honey—neither ever feel they really have enough. One begins with a pair of hives, and during his first year discovers avenues of investigation and areas of study which are possible only if he has more bees, lots more. Or, more prosaically, after his first honey harvest he is so patronized and flattered respecting the virtues of his honey that he decides he simply must enlarge his production to feed the superior taste his friends have suddenly developed from sampling his honey. He may even be tainted by greed and commercialism—how easy it would be, he dreams, to make a substantial additional income by slightly enlarging the already pleasant task at hand with caring for a small number of additional colonies. The reasons for desiring increase—the *rationalizations,* often—are multiple, but the real reason is probably no more specific or complex than that he simply wants more bees than he has.

The most dependable way of getting new colonies on demand—at a convenient time, etc.—is to order packaged bees as you did in the beginning. However, it is occasionally possible to buy whole, established colonies. In the classified adds in the bee journals there are always entire apiaries being advertised for sale. But you are not going to get a whole apiary, or even a single established hive, by parcel post! You may, however, find something within convenient driving distance, and may want to make your increase (or initial purchase for that matter) in this way, but it isn't too likely a practical possibility.

There are other ways to enlarge your holdings which entail no other direct cost than that of the hives and their furniture. Increasing your colonies "at no expense" is always a pleasant prospect, as well as a source of experience in beekeeping everyone ought to have sometime or other. The classic means of such an increase is to capture a swarm, one from your own colonies or a "wild" swarm. Once your existence as a practicing beekeeper becomes known (the intelligence has a way of spreading), a surprising number of calls for you to collect an unwanted swarm that has settled in some one else's living space will reach you. You may also report your willingness to remove and capture swams to your local fire department, or perhaps your local police. At least all of the other books advise you to do this, and there would seem no harm in it. It is more likely, however, that your initial experience with hiving a natural swarm will occur in retrieving a swarm cast by one of your own colonies, in spite of the fact that you have diligently observed techniques of swarm control.

In the old days of skep beekeeping swarm control meant not swarm prevention as it often does today, but swarm stimulation. Certain races of bees were prized for their swarming proclivities, just as races or strains of bees today are prized for their restraint. To practice swarm control either as swarm prevention or swarm capture, something about the mechanics of swarming should be understood.

What precisely triggers the impulse to swarm is unknown; whatever the mechanism, however, the colony begins its preparations for swarming by making queen cells. Bees also make queen cells in order to replace a failing queen, to "supersede" her, but swarm cells are generally distinguishable from supersedure cells. Supersedure cells are usually made somewhere within the perimeter of the comb; swarm cells are usually placed along the bottom or lower edges of frames of brood. Usually the swarm cells are detectably smaller than the supersedure cells, and usually a great many of them are made at one time. The larger supersedure cells are in many ways more elegantly structured, besides being larger. Usually the queen to be superseded is induced to lay in the supersedure cell, even though she is sealing her death warrant. Swarm cells are seldom used by the queen; the workers collect eggs, or even just-hatched larvae, and place them in the bottom of the cell floating in a pool of royal jelly. In either case the larvae are given the same special treatment described elsewhere, which results in the profound difference between the sexually incomplete workers, and their ostentatiously fecund sister the queen.

There is some difference of opinion respecting exactly the sequence of events following the capping of the queen cell. Without trying to be definitive, about three days before the new queen is to hatch, the swarm will issue. The first swarm which issues is called a *prime* swarm. This will be headed by the old queen, and the colony will be left temporarily queenless, awaiting the hatching and mating of a new queen. When this queen hatches, usually her first official act as queen will be to destroy all remaining queen cells; if two or more queens hatch at the same time, sometimes they will fight, with the winner inheriting the rights to the colony. If more than one queen is *allowed* to survive, by the bees restraining the newly hatched queen from destroying the other cells or killing the other hatched queen, the hive will most likely swarm again. These subsequent swarms are called "casts" and several may emerge; the first cast will emerge about five days to a week after the prime swarm. Such casts are usually so small that they are hardly worth catching, but of course are a serious, even fatal, drain on the mother colony. Fortunately they are not too common, although certainly not unusual.

In the meantime, while the new queen is maturing in her cell, certain of the workers have begun their preparations for swarming. Actually there's not a great deal to do other than to "decide" who is to go. The swarm will consist exclusively of young bees, those which have graduated to flight status but have not yet taken up field labors, together with the old queen. Depending on the initial size of the mother colony, the swarm may consist of from one or two up to six or eight pounds of bees. Usually the swarm takes flight in the afternoon, although it may leave anytime after mid-morning. Throughout the morning before departure—sometimes for the previous day or two—the front of the hive will collect a large mass of idle bees. Just before the actual take-off the entire front of the hive will be masked. The exodus of the queen is the signal for departure.

It is an awesome and noble sight to watch a swarm leave a hive. Suddenly the air is filled—*filled*—with bees, forming a cloud in front of the hive, gradually separating from and moving away from the hive as the last bees take to the air, gaining a little altitude until the lower edge of the cloud is five or six feet above the ground. It is difficult to describe the appearance of a swarm on the wing—about like trying to describe the shape and appearance of a drink of water. It's there, a dense cloud, emitting a nervous and somewhat excited hum.

The initial flight of the swarm is usually quite brief—often only a matter of yards if a suitable retreat is found quickly. Bees prefer to settle on a hanging branch of a tree, but can settle anywhere—on a bush, on a wall, on a light standard, a car, anywhere. Folklore is filled with techniques of making a swarm on the wing to settle: beating pans, blowing horns, throwing water, flinging dirt, etc., usually to the recital of appropriate charms. There survives in the Anglo-Saxon language of tenth-century England a charm for catching a flying swarm, together with instruction at certain points in the recital to fling dust taken with your right hand from under your left heel into the mass of flying bees. The technique will work without reciting the charm, by the way. Throwing dust into a swarm will usually induce them to settle quickly on the nearest thing handy. Sometimes, if the queen is forced down, they will form a cluster right on the ground.

Assuming they have not been disturbed by the recitation

of charms, after a brief flight the bees will settle wherever the queen first alights. If this is a first, or "prime," swarm, the queen will be the matriarch of the hive from which they have departed, usually a year or more old. Even though before swarming the bees slim her down by restricting her diet (and her egg laying), she is still somewhat matronly shaped for extended flight. And out of practice: the last time she tried her wings was on her wedding flight. The queen therefore will be somewhat indisposed for extended flying, and will settle wherever her strength or will gives out. The rest of the swarm will alight almost immediately, and form a cluster about her, something like the shape of a classic hornet's nest, a foot or more thick and perhaps two or more feet in length.

Very soon after clustering "scouts" are sent out from the swarm to search for a permanent habitation. There is much discussion and controversy respecting exactly how the search is carried on and the results communicated to the rest of the swarm, particularly how a decision is made by the swarm as to which location to choose, since several opinions might be provided. A version of the "bee dance" is probably the means of communication, returning scouts offering their information of distance and direction by the peculiar geometry of the dance, and measuring its suitability by their enthusiasm. The swarm "chooses" from whatever alternatives are offered it, sometimes within hours, sometimes after a day or more. Rarely—because no suitable place is found, or because two or more alternatives are so evenly balanced in their attractiveness that a consensus cannot be reached—the swarm will never move, but begin making combs and rearing brood where they first alighted. One will occasionally spot such colonies in the woods, or even on a suburban shade tree. Almost invariably winter kills these exposed colonies. Sometimes the swarm will settle in a location without sufficient protected space to make room for the expansion of a developing colony, and will move part of their operation outside by building exposed combs. The future of such a colony is likewise grim.

When, as most usual, the colony selects a site for a new nest, the queen will separate from the cluster and take wing, with the rest of the swarm following and surrounding her. The flight to the permanent site is perhaps more direct and rapid than the initial departure from the hive. When the swarm reaches its destination the workers immediately begin drawing comb, and the queen will begin laying as soon as there is space available. Since each bee has filled itself with honey before leaving the parent colony, and since the bees are all young and in their most efficient wax-producing period, considerable amount of comb can be produced very quickly. Of course the bees will also commence to forage as soon as they have settled in a new location.

Capturing a swarm is usually no more difficult than the mechanical problems whatever the bees have chosen to alight on entail. If it is on a low branch of a tree or on a bush, hiving them should be most simple. As simple as it is, if the beginner can find someone to help who has done it before, he shouldn't stand on pride. Like many simple things, seeing it done is a great deal more instructive than having explained to you how it is done, even with well-meant pictures. The swarms I have hived have mostly been on

Hiving an unusually accommodating swarm of bees

branches, and I have usually proceeded in the most direct way. Take with you an empty hive body, with its entrance blocked (for the time being attach it to its bottom by nailing strips of wood along each side), and place in it three or four frames of foundation. If you have frames of drawn comb, that is even better, although I have always had perfectly good results with foundation alone. Hold the hive body as close to the swarm as possible (which is very easy to say offhand; and usually exasperatingly difficult in the execution), cut the limb carefully, and lower the bees, limb and all, into the hive. Although some of the bees may fly about, for the most part they will hold to the cluster, or return as soon as they discover the queen has been moved into the hive. Now put on the cover, and remove the hive to what will be its permanent location (first blocking the entrance completely, of course).

When hiving a swarm you may use smoke if you wish, but since there is no honey for the bees to be driven to feed on, part of the usefulness of smoke is lost. The bees are already glutted with honey anyway which makes them quiescent, and the whole swarming psychology puts them generally in a non-belligerent frame of mind. You will find that you may physically handle the bees in the cluster without gloves, if you are gentle and take care not to smash any of the bees. Everyone has heard of marauding swarms of bees attacking a human being, to his immeasurable discomfort. It is true that on occasion a nearsighted or imprudent swarm (or rather its queen) has elected a person to cluster on, but such instances are extremely rare. Although it is easy to instruct the proper procedure and frame of mind to the unwilling host in such an instance, nevertheless a person upon whom a swarm has clustered will not be harmed if he remains calm and unexcited. The bees in a natural swarm are looking for a home, not trouble. I just hope if the occasion ever arises I will be able to remember that.

Sometimes it is not possible to cut the branch to collect the swarm which has settled there. In such a case the bees may usually be hived by holding the hive as before, this time with perhaps half its frames in place, and gently stripping the bees off the cluster onto the frames. Once the queen is inside, the bees will be content to remain (usually). You must, however, be quite certain you have collected her. She most likely will be at the very center of the cluster, and it is possible for these last few bees to be overlooked. I have read numerous times that the way to hive a swarm is to reach into the cluster, pick out the queen, place her in the hive, and watch the rest of the swarm follow her in. Lots of luck! I confess I have never tried this stratagem, but while hiving a swarm have thought about it and dismissed the idea as absurd. The other alternatives have always been so much easier.

Another technique I have never tried is to induce the swarm to enter the hive of their own volition, to induce them to think this is exactly the refuge they have sent scouts on dangerous missions so far and wide to discover. They have, after all (most likely) just come from an almost identical hive. While they have nothing particularly in mind at the moment respecting the design of a new residence, if they are presented with a convenient hive, dark and smelling of beeswax, they are very like to seize the opportunity, so the theory goes. Simply (this is usually the most difficult part) place the hive with its entrance abutting the cluster—the hive body resting on a ladder or whatever marvelous engineering expedient you can come up with. Hopefully in a short while the bees will see what a windfall has accrued, and in short order take occupancy. This technique is recommended especially when the swarm is particularly awkwardly located. You can't cut down a telephone pole or hatchet off the eves of the house, or a rain gutter. If you can manage to get the hive in position, this technique is a likely expedient. If you are already in possession of established colonies, try beginning with a frame of sealed brood as bait. One of the overriding instincts of bees at any time is to cover a frame of brood. It is said that in the case of an unusually awkward "lay," you can induce the bees to travel down a length of string to a frame of brood. They will even abandon the queen, according to proponents of this system, in the interest of the young, and the queen herself will be induced in due course to follow.

Once your captured swarm is hived, you should leave them strictly alone for about five days, as you would with a newly established colony of packaged bees. This gives time for the bees to begin drawing comb on the foundation (they do not need to get used to the queen, of course, as in the case of packaged bees), and time for the queen to begin laying eggs. If comb which has already been drawn, as opposed to foundation, is provided, the queen will begin to lay immediately. Therefore, after about five days, open the hive and remove the branch (if you have cut a limb) or any leaves and debris that fell in during the hiving. Possibly the bees have ignored the foundation and begun drawing comb freehand in the empty space (such comb is called *burr* comb). Most likely, however, there will be at worst a few halfhearted bits of burr comb, adhering downward from the cover, or attached to the branch. Remove any such wax, and fill the hive with the remaining frames. If the bees have *completely* ignored the frames (this is quite rare) and begun to hang comb from the hive cover in earnest, and the queen has begun to lay in these combs, you have a problem. I would recommend, in spite of the hazards, removing all of this comb, and making no attempt to salvage the eggs or hatching larvae the queen has already begun. Just take care that in removing this unfortunate false start that you do not also remove the queen. In the unlikely event that the bees have ignored the foundation completely, removing the burr comb will result in a setback in the colony establishing itself, but then life is filled with small contretemps.

It should be unnecessary to feed your captured swarm. Assuming it was a reasonably large bulk of bees, and the season is late spring or early summer, the bees should be able to take care of all the brood the queen lays and keep her provided with drawn comb to lay in. About a week after your first inspection, another inspection is a good idea. You will find, most likely, that a "nest" has been established, occupying three or four frames, depending on the number of bees in your swarms. This will be the center of activity, and will contain the new brood and the freshly laid eggs. If you find there to be no brood (which is highly unlikely) immediately provide them with a frame of brood containing eggs or freshly hatched larvae (optimately, a frame of comb containing a queen cell). From these eggs, or freshly hatched larvae, the colony can make a new queen. Or, if you can be sure of immediate delivery, you may order a caged queen. Most likely, however, this inspection will show everything proceeding well, with capped brood in evidence, and all appearances of a booming economy. In about twenty-one days after hiving, brood should begin to hatch. At this time you should make another inspection, and perhaps begin to induce the colony to enlarge the nest area by systematic replacement of the filled center frames with undrawn foundation from the edges. A very large swarm may well have taken occupancy of the entire hive body. If not, you may leapfrog the inner frames, replacing them with the partially drawn frames on either side of the nest. This process of expansion may be repeated until the bees have been induced to fully draw all the foundation, and work the entire hive body. Once young bees begin to hatch, expansion will of course proceed very rapidly. As soon as evidence shows that the hive body is pretty well occupied, if the season is not too far advanced the colony may be given a super, to store surplus. While you should not raise your hopes too high respecting quantity, it is quite likely that a good swarm, established early enough in the season, may provide a worthwhile surplus before the end of the season. Of course they must first be allowed to establish themselves firmly for their first winter, and it is quite likely you must feed them rather heavily (particularly if the swarm was captured late in the season) to pull them through. There is no reason why the second year they should not be as productive as any of your other swarms.

If the swarm came from your own apiary, it is likely headed by one of your own mature (perhaps marked) queens. If it is a genuinely wild swarm, meaning specifically that it originated from unknown backgrounds, it is possible the colony will turn out to be bad tempered. Whichever is the case, in all likelihood you will want to requeen in the fall. If the origin of the new colony is in any way in doubt, by all means feed them medicated syrup in the fall to prevent the development of disease. (Diseases of bees and their treatment are the subject of another chapter.)

A very easy, if undependable, way to capture swarms from your own colonies which is practiced to some extent by many large apiaries is simply to leave empty hives scattered about the bee range. Reportedly an impressive number of swarms volunteer to take up residence in the abodes thus provided for them. For the hobbiest who maintains only a few hives, however, this practice most likely will not prove very rewarding.

Sometimes, once one's repute as a keeper of bees is bruited about at large, he may be called on to remove an established swarm from a house, hollow tree, or other such inaccessible retreat. My advice is that it isn't worth the trouble, unless you just naturally want to take bows for your mastery of an esoteric art. Bees inside the walls of a house are almost as permanent as the plumbing. The only real way to remove them forever is to disassemble the wall, take out everything, and put the materials back together completely beeproof. If the bees are simply poisoned (a distasteful notion but sometimes necessary) other bees will come around to rob the stores, and eventually it is almost certain to be reinhabited by another vagrant swarm. Also the presence of the wax, honey, etc., will eventually be a nuisance in various ways, such as attracting vermin, or even damaging the structure as it ferments and decays.

There are means, however, by using a wire funnel placed over the entrance (if there is only one entrance, and it is sufficiently restricted) and forcing access into a hive supported in convenient position. Bees vent from their unwanted establishment through the funnel, into the hive, and out the entrance to the hive. They can return to the hive, but may not find their way through the narrow opening of the funnel back into the home nest. After a short time (probably thirty days) most of the bees, including the most recent hatches, will be captured and contained within the new hive, where they will begin working. The hive is then moved, given a queen (and perhaps a frame of brood or two), and the original opening sealed off. As I said, it is hardly worth the trouble. And if removing the bees, rather than the obtaining of a new hive, was the purpose to begin with, the job has still been only half done.

It also sometimes happens that one comes across a swarm some boob has captured and installed in something he has nailed up of planks, or worse, an old apple box probably with a thin piece of plywood for a top. If there is any internal furniture at all in this hive, it will most likely consist of a pair of sticks crossed in the middle for the bees to hang comb from. The only inducement to accepting such a hive ("bee gum" or "box hive" are the general appellations for such makeshifts) is sentimental concern for the welfare of the bees themselves. Like dog and cat fanciers, beekeepers don't like to see the objects of their fancy abused. Only once have I tried to salvage such a colony, and the process was tedious but finally successful. The bees were ensconced in an apple box, and I proceeded by simply placing a hive on top of it filled with foundation and waited until the bees moved up and the queen began laying in it. I then broke up the box one evening and dis-assembled the combs in front of the new hive. With the coming of the night chill the bees crawled into the hive, and early the next morning, before the bees were flying, I carried away and destroyed the black and misshapen combs, brood and all. The rehived colony itself was safe and secure, and strong enough to survive. However, do not go looking for such "windfalls"; if you do inherit one, you will probably have to resort to some such field expedient as I describe.

Not long ago a correspondent to the *American Bee Journal* explained that bees can be induced to give up work-ing on combs by turning them upside down. Since the cells all angle slightly downward, they are useless to the bees if inverted. Therefore in a situation such as I described, it should be possible by turning the "gum" upside down to force the bees immediately to abandon further work in the old comb, even moving any stored honey upward at the same time. Brood already in the cells would be hatched, but the queen would not lay in them again.

Another way of propagating bees, or getting started in the first place, is by ordering bees on the combs, with an established and laying queen, rather than buying them in packages. Such abbreviated colonies—called "nucs"—are ad-vertised in the bee journals and some of the catalogues.

Really, however, they are not for the hobbiest. Practically speaking the only way for most of us to obtain such nucs is to pick them up at the breeder's. This is usually practiced only for a large operator, or for one fortunate enough to just happen to live conveniently near such a producer. There are also many restrictions regulating interstate transportation of bees on the comb, enforced to combat the spread of disease.

A curious footnote to history concerns the early-days transportation of bees to the western states. There was actually a "bee rush" to California hot on the heels of the gold rush. Before the agricultural potential of California's Central Valley could be appreciated, bees for pollinization were necessary (one of the most dramatic demonstrations, by the way, of the crucial relationship of bees to agriculture). Consequently, for a short period in the 1850s bees were one of the most valuable import items. Various means were attempted to transport the colonies to ensure safe arrival. Bees from New England were shipped around the Horn, or transported across the Isthmus of Panama. Crossing Panama the bees were sometimes allowed a breathing spell by situating the hives, opening them, and letting the bees fly for a few days. Colonies were also transported by wagon across the plains and mountains from St. Louis, and incredibly, many survived. The boom was short-lived, but immensely profitable while it lasted until local reproduction could keep up with the demand.

One practical and relatively easy means of increasing your colonies is a procedure known as "splitting." To increase your holdings in this way you must have at least two strong colonies. Depending on your local conditions, it probably should be done in late May, after your spring build-up has put your colonies in top condition, and before the main honey flow begins. Many beekeepers, by the way, practice splitting less in the interest of increasing their colonies than as a measure to help in swarm control, a topic considered more fully elsewhere.

To split your colonies, take four frames of very solid brood from one hive and place them in the center of your new hive surrounded by frames of foundation, or drawn comb, if you have it. Then shake from the brood frames of your second colony a goodly quantity of bees onto the frames of the new hive (inspect carefully to make certain you do not inadvertently shake off the queen). You should then immediately situate the new hive where it is to remain permanently, using a restricted opening, further restricted by closing it off with a small bunch of grass (this will discourage robbing, and the inside bees will readily remove it as the grass withers). You may then introduce a caged queen, using the same technique you did when installing a packaged swarm. The process is best done in late afternoon. You may even proceed without using a queen, provided that there is either an occupied queen cell among the brood you introduce, or unhatched eggs (or larvae no more than twenty-four hours old). It is much better for a number of reasons to introduce a mated queen. If you are running a shipshape apiary and practicing queen control, you will want to introduce a "blooded" queen later anyway.

The disadvantages of splitting should be obvious—it reduces the strength of your established colonies very near to the time when their work will be to your greatest advantage. If you split only strong hives, however, the weakness is probably more than offset by the reduced likelihood of the colony swarming that year. The advantages of splitting are that it does work. Most of the bees you have introduced are young—like packaged bees—and have not yet oriented themselves to their parent hive. Consequently they will accept their new hive readily. Older bees that have inadvertently been included in the transfer will return to their parent hive in short order. Putting the grass in the opening temporarily restricts movement; if the bees have not pulled it apart, it should be removed in a couple of days. Since the bees transferred are "hive bees," too young yet to work in the field, you will see very little activity around the front of the hive for the first few days. Once the transfer has been made, it is advisable to leave the new colony strictly alone for several days; treat it in most respects after installation like a newly hived swarm of packaged bees.

A rarer means of increasing your holdings is simply to fall heir to one or more hives of bees. Particularly in a rural location, someone will not or cannot move his bees when he leaves for the city. Often one falls heir to bees when the member of a family who tended the bees dies, and the legitimate heirs have no interest in their legacy and seek you out. Particularly in this latter exigency, more often than not you will simply be offered the gift of a hive or two, and probably some equipment. This is an occasion when you absolutely must look a gift horse in the mouth. Old bee equipment, in the first place, just isn't worth the risk you run of contaminating your own healthy hives with disease organisms. Old hives in which the bees have died out present almost prima-facie evidence of disease. You must not talk to the giver in these terms, of course; to suggest that their bees might have been diseased produces much the same emotional reaction as suggesting their own persons to be venerally infected. Such gifts of equipment are best accepted graciously and then privately burned. It's the safest route, and serves a good if unheralded service to the beekeeping fraternity at large.

Hives actually containing bees should be examined carefully before accepting. If the hives have been regularly checked by the state bee inspector, they may be considered safe, but often they will not have been on the inspector's list. The law in most states requires bees to be registered in order that they be inspected for disease, but a surprising number miss ever appearing on the rolls. If such bees have been neglected for a while, which is usually the case, the swarm likely will be weak regardless of their general state of health. Building them up to a strength sufficient to winter may be difficult, and probably not worth the effort other than as a project you engage in for the fun of it. Only accept such bees after checking thoroughly for disease, and if there is any doubt do not risk contaminating your own apiary with infection.

10

Honey and Its Sources

IT IS PERHAPS SILLY to begin at this late date to explain what honey is. Most people can identify it readily. However, in terms of composition, its sweetening agents are fruit sugar (levulose) about thirty-nine per cent, and grape sugar (dextrose), about thirty-four per cent, with the remainder of the composition being water and other trace substances. Chemically the sugars which comprise honey are quite different from cane sugar (sucrose), and while bees readily take up syrup made from cane sugar which is provided them for wintering or the spring build-up, the presence of cane sugar is considered an adulterant in commercial honey, and stringently prohibited. And the presence of sucrose is rather easily detected by routine spectrographic analysis. Besides its sugars, the largest single constituent of honey is of course water. The quantity of water allowable in commercial honey is twenty per cent. This does not mean that honey producers would otherwise dilute their produce; presence of honey in amounts above that figure means in practical terms that the honey was inadequately cured before extracting. Such a condition can occur if the honey is extracted before the majority of cells have been cured to the bee's satisfaction and capped. Such honey tends to ferment, and its general flavor and quality is substandard.

In addition to its major ingredients, honey contains in varying amounts pollen, volatile oils, gums, formic acid, albumen, digested fat, and traces of other substances. It is in these lesser substances and the esters from the flowers that many of the more romantic properties attributed to honey are said to reside—particularly the fabled medicinal and curative properties, as distinct from the obvious nutritive values.

This is no place to either endorse or refute such curative powers. It is certain that honey is a source of quick energy, and a spoonful or so will consequently give you a feeling of well-being if you are tired. What may prove to be honey's most practical short-term effect, however, as indicated by recent experiments, is that it has value in control of hangovers, the sort resulting from excessive alcoholic indulgence. It is even said to have a sobering effect, and in such countries as England and Australia where driving under the influence is heavily penalized, tipplers have taken to medicating themselves with a couple of spoonfuls of honey to sober up before mounting their machines. According to reports, this treatment works both physically and mechanically, even convincing the instruments of the law which detect the degree of alcohol in the blood stream that the donor is sober and in full possession. Don't take my word for it; I merely pass on what I have heard.

I will not repeat the perhaps too familiar attributes otherwise claimed for honey as an agent of analgesis and cure. It is most likely inevitable that honey should have as many mysterious, even mystical, properties attributed to it, coming as it does from flowers. To state the obvious, honey is basically the secretion of the nectaries of flowers, reduced down and otherwise processed by the bees. As it is first gathered, the nectar is thin, sweetish, generally very watery. It is collected by the bees with their tongues, and stored in their "honey stomachs" for transportation back to the hive, and in this stomach undergoes certain changes by action of enzymes provided by the bee's body. Once inside the hive the honey is deposited more or less temporarily in a convenient cell—sometimes a cell which is still abuilding. The nectar is further processed in handling by the "hive bees," who spread the nectar over large areas of comb to facilitate evaporation. The nectar must be reduced in volume by as much as seventy-five per cent before the processing is completed, and most likely it will be handled many times and acted upon by many honey stomachs before the bees are satisfied it is pure honey and suitable to be bonded and capped. During a good honey flow, you may stand beside a hive during a warm night and hear a rather loud humming as the bees process and evaporate the nectar, ventilating the hive with vigor as part of the process to replenish the moisture-laden air with fresh air from outside the hive.

If you are watching your bees closely, you will notice frames filling up with honey but remaining uncapped for what might seem to be extended periods. If you taste a bit of this uncapped honey you will most likely find it not completely palatable. The flavor tends to be sharp, even acrid. The difference between this honey and cured honey is about the same as between fresh wine and aged wine. Even after it is capped honey continues to improve in flavor and mildness over a considerable period of time in quite the same way wine does. In fact, honey made from certain plants is said to be most unpalatable until it has aged several months.

While the reputation of honey for its wholesomeness is rivaled perhaps only by milk, there are persistent traditions of poisonous honey. One of the earliest surviving reports, from Xenophon's account of the retreat of the Ten Thousand in 40 B.C., describes sickness from eating combs of honey collected by the marchers. Modern authorities believe this honey was from Pontis Rhododendron. Honey from these blossoms has a toxic quality which seems to be present only while the honey is still uncapped, and eating it produces vomiting, disorientation, and even in isolated cases, death. Mountain laurel seems also to have poisonous properties, but also only if the honey is eaten uncapped. Both of these honey sources are sufficiently rare that even in a locality where the plants exist the possibility of the bees' collecting a surplus is very remote, and the probability of eating this honey uncapped even remoter. It is persistently reported that bees collecting pollen from poppies display symptoms of grogginess, and seem to have difficulty locating the entrance to the hive. I have never read anything that sounded like an authoritative study confirming this phenomenon, and considering the usual process by which crude opium is extracted from poppies, this account sounds mostly fanciful.

A source of "honey" which constitutes a real trial for

beekeepers in some locales is "honeydew." This is not properly a honey at all, since it is not collected as nectar from blossoms. Rather, it is processed from the sticky, sweetish sap exuded by many varieties of trees when they are attacked by aphids and other similar sucking insects. This excrescence is the familiar sap which in some areas plasters the windows and paint of parked cars, or even makes the sidewalks slippery. Bees usually gather this secretion only if normal nectar sources fail, during a period of hot, dry weather, for example. The flavor and quality of honeydew varies with the source and location, but most beekeepers consider it worthless. Although it has some commercial uses, it is generally unpalatable, and if bees are allowed to retain it for winter stores, it frequently causes dysentery (because of its high protein content) and can even result in a hive dying out during the winter when extended bad weather keeps the bees from taking "cleansing flights," as trips to that great potty in the sky are called.

In some European countries, however, there is an actual preference for honeydew, depending on its source. Germany, for example, consumes much honeydew, mostly collected from pine trees. Under certain circumstances pine trees exude a sap, locally called "pine sugar," which bees collect and store. Pine sugar is common in some parts of the Northwest, especially during periods of extreme heat. Honeydew collected from this pine sugar is reported to be quite palatable. For the most part, however, there is little or no demand for honeydew, and beekeepers generally deplore its collection.

During periods of scarcity, bees will often be observed collecting juice from split fruit in an orchard, or particularly split grapes in a vineyard. In spite of a folk-reputation to the contrary, the bees do not themselves split the fruit (their mandibles are completely inadequate for the task), but visit only fruit which has been caused to split from other causes—rain, insects, birds, etc. The juice thus collected is, like the honeydew, of little or no value. Most likely it will simply ferment in the cells, and spoil the comb in which it is stored.

Except for honeydew, which is a questionable exception at best, the total dietary requirement of the honeybee is provided by flowers. From the nectar they get carbohydrates, from the pollen protein. Although most flowers provide some of each substance, this is not uniformly true, nor are flowers at all equal in providing these necessities. Assuming your colonies to have adequate stores of honey, or that they are supplied syrup supplement, their most compelling spring need will be for pollen to provide the protein necessary for spring brood rearing. If you feed pollen substitute to provide early spring protein your bees will abandon this in favor of natural pollen as soon as the earliest blossom sources become available.

The mechanics of blossom construction separates flowers, at least from the point of view of the beekeeper's interest, into three broad categories—those depending primarily upon the wind for pollinization, those depending upon moths, and those depending upon bees and their kindred. The pollen of plants depending upon the wind for fertilization tends to be very light and dry, and consequently it is most difficult for bees to handle efficiently; in fact, much of it is impossible to pack into the pollen baskets. Since these plants do not require the visit of insects, they secrete no nectar to entice them. Nevertheless, if other sources are in extremely short supply, bees sometimes make the effort to collect from these plants what pollen they may. Many trees are of this sort, for example walnuts and elms; most grasses, including corn and the other cereal grains, are also wind pollinized.

Those flowers which are designed primarily for fertilization by moths and butterflies—the *lepidipteroid*—usually secrete their sweetness in nectaries that are too deep to be reached by the tongues of bees. Some such plants become highly specialized in their requirements for fertilization. The yucca, for example, requires the specific attention of the yucca moth, which lays its eggs in the ovaries of the blossom, fertilizing the flower in the process. When the larva hatches it eats *most* of the seeds, but without the yucca moth there would be no seeds at all. The evening primrose, and most lilies, are lepidipteroid. One of the pleasures of growing evening primroses is to watch the large, almost bird-sized sphinx moth which visits the blossoms as soon as they begin to open in the evening. Such plants frequently produce large amounts of pollen, and bees sometimes work the blossoms even though they have little or no nectar which the bee is able to reach. However, at the same time these blossoms are in evidence usually there are large numbers of other blossoms which have greater attraction to the bees.

Those blossoms less selective than the lepidoptera are further subdivided into those which cater to honeybees and those which cater to other insects. Some flowers are totally promiscuous and unselective. Bees, flies, birds, even beetles indiscriminately may scatter their pollen. Other flowers require the servicing of insects with extraordinarily long tongues or proboscis, or of particular strength. Some of the legumes fall into this latter category. The blossom of the alfalfa, for example, is so constructed that an internal mechanism of the blossom must be "tripped" before the pollen is released. The honeybee is not strong enough to provide this service; only varieties of the solitary bees or bumblebees —particularly one called the *alkali bee*—are capable. The use of pesticides in many seed-producing areas of the west nearly eradicated these necessary insects, until their importance was realized and steps were taken to preserve and perpetuate the species.

The different kinds of clover vary widely in their availability to honeybees. The most important clover, and probably (commercially speaking) the most important of all honey plants, is the white clover. It is said that this clover has followed the spread of the Anglo-Saxon races, and where the Anglo-Saxons go, it becomes the dominant honey plant. This appears to be verified in the North Americas and in Australia and New Zealand. The common white clover is frequently grown for animal pasture, is found in lawns, and thrives in suitable waste places such as fence rows, road banks, etc. Its small, white, globular blossom is profuse in its yield of nectar, and its size is ideally suited for the management by honeybees. Each tube in the multiflora head of its blossom contains nectar, so despite its apparent dwarfishness, it is heavy in total yield of nectar.

Other clovers, indeed most legumes, are at least potentially large yielders of honey. Red clover, perhaps the most spectacular of all clovers (an "improved" variety is now cultivated as a garden flower), is a good illustration of the mechanical limitations a blossom may impose on its importance as a honey plant. There is a German folk saying that the Lord prohibited the bee's use of this lush flower because it broke the Sabbath. At any rate, while its blossom consists of a multiflora head like the white clover, the tubes to the nectaries are crucially longer than the tubes of the white clover; the maximum "reach" of a bee's tongue is about 7.9 millimeters, while the average depth of the red clover tube is about 9.6 millimeters. There is just no way to stand tippy-toe when reaching with the tongue. Therefore, while the honeybee may visit the red clover for pollen, ordinarily she has little chance of profiting from the otherwise rich nectar the blossom offers. If conditions are especially optimum, as for example after a quite heavy dew, the tubes may be sufficiently filled that the bees' tongues may reach the nectar—in which case, because of capillary action, the bee may actually succeed in draining the tube. Practically speaking, however, the only future for the red clover as a honey plant is to breed either clover with shorter nectaries, or bees with longer tongues.

The various other small or low-growing clovers are also quite important for bee pasturage. The taller, rank sweet clover is especially important. In areas where it commands the fence rows and roadside ditches, it may be a major nectar source. All of the clovers provide honey of virtually indistinguishable character. Typically it is light colored or water white, very mild and very pleasantly flavored. Most clover honey, including alfalfa, crystallizes rather readily, producing a solid form of honey which is quite white and quite fine-grained.

Besides the clovers, a wide variety of blossoms are important to the internal economy of the hive, but seldom produce surplus honey. Fruit blossoms, for example, are a major early spring source of nectar, but the nectar harvest (along with the pollen collected) is utilized almost immediately in the rearing of brood. By careful manipulation—supplemental feeding, principally—it is possible to secure and sequester away a small amount of honey from the fruit blossoms, but this is seldom done on a commercial scale. Even so, it is difficult to obtain anything like a pure fruit-blossom honey, since so many other nectar plants may be blooming at the same time, many of which the bees actually prefer to the fruit blossoms.

And while early spring bulbs—daffodils, narcissus, snowdrops, and the like—provide an important early source of pollen and sometimes nectar, it is virtually impossible to obtain any surplus from such sources. This is most likely just as well; in all probability the honey from such sources would be unwholesome for human consumption, perhaps even toxic. Similarly, while dandelions bloom early and are prodigious producers of pollen, as well as providing nectar quite freely, most of this yield is consumed in the spring build-up of the hive. Honey from dandelion is reported to be quite badly flavored, and most likely it is fortunate that circumstances do not favor its storage as surplus.

Another very valuable early spring source of pollen, and

some nectar as well, is the pussy willow, which appears among the earliest of blossoms in most northerly climate zones. It is not necessary for the tree to produce the elaborate catkins of the French pussy willow for the tree to be a lavish producer of pollen; all forms of the willow produce heavily, and are invaluable for spring build-up. Willows are so easy to grow (too easy, perhaps, by most standards) that if such trees are not otherwise available within flight range a beekeeper might well think of establishing some near his whereabouts.

The hazel and birch are similarly heavy producers of early spring pollen, and warrant consideration as plantings adjacent to one's apiaries. Like the willow, both set out catkins in the early spring, upon which the festoons of pollen are clearly visible even at some distance. The cottonwood, where it occurs, is similarly a good early source of large quantities of pollen.

Of the garden plants, one of the earliest yielders of pollen is the crocus. Usually border plantings of this attractive plant are sufficient to make it a significant and worthwhile source to your bees. In the past, in fact, so great is its attractiveness to early foraging bees that beekeepers used to "charge" their blossoms with pollen substitute for the bees to collect. Such substances as pea flour, soy flour, even wheat flour were sprinkled into the blossoms, as well as commercial blends of pollen substitute. Such disadvantages as the tendency of these substances, thus administered, to become pelletized in the combs and become useless to the bees has tended to make this practice of supercharging the crocuses in the spring fall into disuse. And of course the techniques of administering pollen substitute mixed with honey as a paste inside the hive is more convenient and certainly more dignified than dusting about among the crocus blossoms with ersatz pollen.

While clover is the queen of the honey producers, other varieties of blossom are of course extremely important, and in some areas comprise the dominant source of surplus honey, particularly of specialty honeys. By careful manipulation, for example, wild raspberry honey may be produced in surplus quantity in some areas, particularly upper Michigan, and it has a reputation of elegant flavor. On the other end of the continent, tupelo honey is the specialty of low-lying southern states (from the tupelo tree), particularly Florida. Tupelo honey is characterized, aside from its unique and mapley-flavor, by never granulating. It is sometimes mixed with other honeys as an agent to retard crystallization.

One of the less immediately obvious features of honey-producing plants is that frequently their characteristics vary, depending upon where they are grown. Too many variables are involved to always (or *ever*) know why this is true—climate, moisture, soil conditions, etc. What may be a good honey producer in one region may be an insignificant source in another; or the whole character of the honey may change. So the black locust, a true native to North America, is in its homeland considered an indifferent if moderately important honey source, even where it occurs in quantity. In Central and Eastern Europe, however, where it is naturalized, it has become a honey plant of considerable importance. Under a naturalized name of "false acacia" it is the source of a major variety of specialty honey, known for its delicate, almost spicelike flavor.

The extensive varieties of honey-producing trees are no doubt of considerably greater importance to the urban or suburban beekeeper than his more lavishly supplied rural compeer. Such common shade trees as maple, bass wood, the various tulip trees, and locusts can provide an important surplus. In England the plane tree, or lime as it is called, is responsible for a considerable percentage of the total national honey crop. In Australia the eucalyptus tree is depended upon as an important honey producer.

Speaking of honey plants raises questions about the possibility or advisability of planting flowers suitable for bee forage. On a large scale, this is not economically feasible, unless the plants can have additional use other than bee pasturage. It is possible to plant otherwise waste spots with plants the bees like, particularly hardy clovers (sweet clover is ideally suited for many such areas). I myself have made numerous surreptitious plantings of sweet clover along roadsides within the range of my hives, and considered that I beautified the country at the same time. If convenient, a pussy willow or clump of willows will be worth planting solely for their value in the spring build-up. Mr. Hawes, in his book on English honey plants, exhorts the reader to pressure city planners to select shade trees with an eye to their usefulness to urban beekeepers. He also suggests that freeways would have some of their ugliness corrected if they were bordered by trees—varieties with bee-appeal, of course. This idea is not a bad one on either count. Certainly the dreary freeways of this country could be improved by extensive borders of trees. And of course the appearance of any tree is improved by the presence of happy and industrious bees

11

Odds and Ends and Afterthoughts

A NUMBER OF TOPICS with more justification to idle curiosity than real practical concern have been omitted from the preceding chapters in the interest of economy and coherence. Some topics with practical value have been omitted simply because there didn't seem an occasion to discuss them. The author has also neglected to offer a few quarrelsome and querulous comments of his own, which seem to be *de rigueur* for all manuals of beekeeping. These matters, practicalities as well as curiosities, will be attended to here as a sort of general addendum, presupposing that in the interim if the reader has not actually confronted bees as a practical reality, he has meditated on them sufficiently that at least vicariously they have assumed the dimensions of practical reality.

So to a practical matter. I trust the reader has either searched for a queen in a live colony by now, or at least on reflection has wondered how one goes about finding her. In most books on beekeeping the authors either state directly or at least positively intimate that they can at will go into the apiary and find a queen as easily as they can find the hive. I confess I can't. I know how to look for the queen, but I still can't say that I can find a given queen at will.

If you have a marked queen, the task is very much simplified. In any case, when you go into a hive to find the queen, do nothing else at that time and look for nothing else until you do find her, or give up. The principle is about the same as hunting for anything else—rabbits, microbes, bibliographic entries. When your eye is educated as to what to look for, there's nothing to the finding. When you search for the queen, do it systematically; you obviously should look for her where it's likely she might be. Obviously you will most likely find her in the brood chamber, where there are eggs or freshly hatched larvae. Begin your inspection by removing a center frame. If it does not contain eggs, it is most likely she will not be there. Set the frame to one side, and examine the next. If it does not contain eggs, move it sufficiently to facilitate the removal of the next frame until you either find eggs or reach the end of the hive. If you find no eggs, go back to the center of the hive where you started and work back toward the opposite wall in the same manner.

When you find a frame containing eggs, begin to search slowly and carefully for the queen. Handle the frames gently and with care. The queen is a shy thing, easily frightened, and if frightened she tends to hide—either amid a throng of bees or under the edge of the comb, or by hastily leaving the comb and racing toward a dark part of the hive. If she has not been disturbed by rough handling of the frames, most likely you will find her surrounded by a group of workers, with their heads all pointed toward her like spokes of a wheel. If you go through the brood nest without finding her, it's best to put everything back together and close the hive for a few hours before beginning the search again.

British books on beekeeping particularly make it sound easy to find the queen, and intimate that anyone who can't is a particular kind of klutz. These books also intimate that you should look the queen up every time you get a chance, or at least every time you have business in opening up the hive. My feeling is that when you open a hive you do it for a purpose, and your purpose is best served by disturbing the colony as little and for as short a time as possible. I would think it certainly not in the best interest of your colonies' production and morale to conduct a queen search unless it serves a particular purpose. You can tell if you have a queen, and how she is performing, by looking at the frames of the brood. Looking at the queen herself will tell you nothing but what color she is, unless you are a skilled judge of queens.

An amateur beekeeper in quest of improving his science or craft will undoubtedly find a number of novel and curious notions and practices in reading of bee culture as practiced by the British—described in the "bee press," as British authors like to call it. These ideas and practices may startle and even discourage an amateur, especially if he comes to them totally innocent of practical beekeeping. For example, the variety in the design of British hives is a source of wonder, if not incredulity. Admittedly the patterns of British hives have charm and quaintness, even elegance, which the standard American hive may lack. But the only book

I have ever read by a professional British beekeeper scorns those elegant hives, and attributes to their clumsiness, complexity, and expense a major reason why England produces so much less honey than it consumes, and apparently so much less honey than the flora of the country is capable of producing. In his business he uses standard American equipment (modifying the bottom of extracting frames), and advises its use by anyone seriously interested in keeping bees, certainly for anyone interested in keeping bees for a living.

The British amateur also writes at great length about the various "manipulations" required in keeping bees. The term is so handy that I find I have picked it up, but by *manipulations* (always plural) the British amateur intends such a variety and complexity of impositions on the colony of bees that one wonders that the native British black bee must have accepted extinction with gratitude, perhaps even soliciting the acarine infestation as a kind of suicide. Again, the one professional British beekeeper I have read scorns "manipulations," on the one hand doubting if any amateurs actually do all of those things to their bees, on the other hand attributing at least the intent to manipulate with its own share in shortchanging British honey productions.

As Thoreau said, *simplify*. Except for considerations imposed by using hives with movable comb, do as little to your bees as possible, and vary conditions from what they would be in the wild state only by clearly improving conditions of their natural life cycle—providing them with unlimited room, keeping them headed by a young queen, feeding them when the stores run low. But don't pester them. In all honesty, I think every setback my bees have ever suffered has in one way or another been directly or indirectly the result of my own bumbling or pestiferation. You must always learn to live with the bees; it is arrogant if not sacrilegious to meditate making them over in your own image.

To precipitate toward another topic altogether, some people are attracted to practical beekeeping as the only way to be assured of pure, untampered honey. The practice will do that, of course. Early in this century a wild rumor of contamination nearly ruined the beekeeping business. It was rumored, even "confirmed" by government bureaus, that ersatz honey, made of sugar syrup, was being sold in "comb" pressed of paraffin. For all its patent absurdity, the rumor was amazingly long-lived (and I suspect there are those alive who still believe it). Rumors still have currency about contamination of honey, particularly of damaging the honey by improper handling. Food faddists are particularly keen on the notion that marketed honey is polluted, and pay premium prices for "raw" or "natural" honey. From my cursory experience with such "premium" honey of the health food stores traffic, unscrupulous (or pragmatic) honey producers bottle up honey which is unmarketable because of its dark, strong flavor, and making a virtue of a necessity call it *natural.* Which there is no question it is! My experience has shown, both as a child and an adult, that there is some sort of mysterious bond between *healthful* and unpalatable.

While on the subject of that which is at least peripherally fanciful, another product of the hive needs to be discussed. The propolis, which certain varieties of bees collect to the extent of absolute nuisance and all bees collect to some extent, is a substance about which there lingers a certain mystery. Presumably this substance is collected by the bees from sap or pitch exuded by buds, although there is not even universal agreement about this. The bees use propolis in the hive to fasten loose bits of hive furniture securely together, to the immense and unvaried annoyance of the beekeeper. They also use it in the fall to seal off or restrict the entrance of the hive (hence the origin of the substance's name, *propolis*, "before the city"). Any substance disagreeable to the bees' smell or counter to their sense of hygiene is often liberally dabbed or effectively varnished with the substance.

This suggests one of the curious properties of propolis. It seems to be an antibiotic or perhaps a fungicide of some power, and in this capacity bees use it to cover and neutralize wood contaminated with organisms causing disease, or potential sources of infection within the hive. They also use propolis to encase and in effect mummify carcasses of mice, etc., which die or are killed inside the colony.

Propolis was used as a surgical dressing by the British during the Boer War, and its history as a medical unguent is rather ancient. In very recent times medical research has been systematically conducting investigations of its potential as an antibiotic. It is not impossible it may eventually be added to the modern pharmacopoeia, in which case propolis might well become in the future a product of the hive as valued as it is now despised.

One of the most romantic traditions having to do with propolis identifies it as the wax or varnish used by such master Italian violin makers of the seventeenth century as Stradivarius for finishing their marvelous instruments. If this tradition is true it should be easily verified through modern techniques of qualitative analysis; I simply pass it on as a happy tradition. As beautifully as propolis varnishes a piece of wood inside a colony of bees, one would not be surprised to find it had furnished the finishes of the world's greatest violins.

Another byway of bee lore, this time clearly in the tradition of superstition, *explains* a superstition by associating propolis and varnish. There is a country tradition that when a member of a family who keeps bees dies, the bees must all be "told" of his passing, or they will abscond. There is a pleasantly melancholy poem by Emily Dickinson about the practice titled "Telling the Bees." A number of authors mentioning this old superstition repeat the same explanation for the origins of the tradition. They like to say that the tradition must have got its start because bees would visit a freshly varnished coffin in the laying-out room in search of material to use for propolis. I like to locate the origins of traditions as well as most (I read to my children from a copy of *The Annotated Mother Goose*), but this explanation seems most fanciful. It is so unlikely that a diseased beekeeper would be encased in a freshly varnished coffin in the time this tradition must have originated that clearly another source of the superstition must exist. Like most rational people, I do not like to have my superstitions tampered with ineptly. Until such time as a better explanation is offered, I suggest the practice be continued.

Occasionally in the preceding chapters the *morale* of the

hive has been alluded to. Although agreement is not universal, there distinctly appears to be a condition of spirit within a hive which is reflected by the general willingness of the hive to work and its productivity which may best be designated *morale.* Probably no situation of higher morale will be found than that of a newly hived prime, natural swarm of bees in early June. In part their activity may be attributed to the fact that the swarm consists entirely of young bees, in their most versatile and productive time of life. Nevertheless, the amazing production of drawn comb, flower products, and general movement to expansion cannot be accounted for by one single cause. It is an impulse which carries through the entire season, beyond the lifetime of the bees originally hived, and a swarm hived in June will frequently astonish you with the surplus they house before the end of the honey season. Similarly, a colony that has been headed by a weak or failing queen turns from a psychology of survival to one of aggressive expansion upon the introduction of a new young queen. Conversely, colonies weakened by wax moths, disease, or generally crummy living conditions tend to be lethargic, lackluster, even lazy. Cleaning up a neglected hive will almost invariably impel it into a frenzy of work. Occasionally a colony which in all appearances should be in good shape just doesn't seem to be up to much; the only thing you can say frequently is that its morale is down. Usually such a lackluster colony can be restored to pride and productivity by removing its queen (even if it's a young and fecund one), leaving the colony queenless a day or two, and introducing a new queen. Somewhat like people, apparently bees never miss the water till the well goes dry. And there's no dryness for a colony of bees like being without a queen.

In the late summer, when the robbing danger is highest, it is the hives with low morale, as much as the physically weak hives, that become victims of robbing. Robbing itself seems a matter of mysterious psychology too, if the term is not misused applied to the insect world. The robbing fever seems to be stimulated by dearth of nectar, and is completely unrelated to the stores a colony may already have put by for winter. Indeed, it is those whose stores are marginal or inadequate who are the normal victims of robbing. It is hard to talk about robbing without becoming at least partially anthropomorphic. Some of the things that occur in an apiary under the robbing fever probably will never be believed until a skeptic has seen for himself. Sometimes robbing will occur as a mob phenomenon; hordes of bees from one hive will break into another hive, which lacks either ability or will to resist, and without preliminaries or finesse simply remove everything that isn't tied down. More usually, robbing will be restricted to a few more or less professional individual robbers. These bees actually *look* evil; through their numerous and skulking intrusions through narrow places into other hives, their body hair is worn away to an extraordinary degree, until they possess an appearance that is positively oligophrenic. In practicing their craft or vocation, they approach the hive to be robbed from the back and sides, searching for cracks they can enter into, avoiding the main line of bee traffic, and assuming a demeanor that can only be called furtive. It is said that once a bee has had the heady experience of ripping off refined honey from the cells of another hive he from that moment on gives over the honest pursuit of nectar and will never again labor in the fields of the Lord. I do not know this for incontrovertible fact, but all of my observations confirm it.

It is absolutely incredible, unless you watch it yourself, to see the deviousness and ingenuity of a robber bee trying to breach the guard of a prospective mark. During the season when robbing becomes a problem, security is tightened on all of the hives, and the line of ventilating bees is replaced by a line of guards, who check every returning bee carefully before it is allowed to enter. During the heavy honey flows you will virtually never see a guard stop a returning bee; when the honey flow has slowed and stopped you will see virtually no bee admitted uninspected. If you observe a given hive for some time, you will quickly come to recognize the robbers, the slick, oily little insinuators! If you didn't loathe them so you would come to love them for their gall and ingenuity. One robber will attempt to fast-talk his way in through sheer crust. The guards will meet him, they will caress each other with their antennae, and the robber will be ferociously expelled. He will fly off a short distance and return by skulking along the side of the hive until he can plop down almost inside the first line of guards, and go through interrogation again. Finally the outside guard will pass him, only to be restrained by another guard just inside the door. He will be dragged away protesting indignantly (robbers never fight seriously, because they can only lose), to return and start the whole thing over again. Clearly in the time elapsed the robber could have made several successful trips to the field, and with far less effort. Robbers will actually resort to stratagems; pretending to carry away a dead bee seems to be a favorite ploy to work themselves inside the guard. One will frequently fall to work helping one of the legitimate inhabitants carry out a corpse, only to drop the load and make a dash for the inside when he "thinks" the guards have lost interest in him. A robber will even collect a small bit of pollen to provide ersatz credentials to convince a skeptical guard that he belongs to that colony and is contributing to its upkeep. While it all looks very trivial and ill worth the while when viewed on an individual basis, if a robber or crew of robbers actually succeeds in infiltrating another hive, the way is made easier for subsequent incursions by larger numbers from their home hive, until eventually the total morale of the hive being robbed is broken and they no longer put up any resistance at all.

But I'm sure you don't believe this unless you've seen it for yourself. After all, the animal world is supposed to teach us morality, not hold up to us the mirror of our own vices. Just remember that the next time you heist a super off a hive you have spent the summer cunningly manipulating.

Annotated Bibliography

THE FOLLOWING is not offered as a comprehensive guide to major books on beekeeping. Rather, it is intended to be a preliminary guide to additional reading on the subject. Some books are selected because they cover the subject of beekeeping comprehensively, technically, or professionally, or because they cover such professional topics as grading, processing, and marketing, etc., of honey, which are beyond the scope of this present book. Others are included because their entire topic is one I have been able to touch on only briefly or partially, such as honey plants, queen rearing, etc. And some books are included purely for their intrinsic readability, or even for their eccentricity. Most, although not all, are more or less readily available through the average or even modest library. In addition, the reader is particularly advised to check the publications available (many of them free) through the Department of Agriculture, Agricultural Extension Service, state agricultural colleges, etc. These publications are frequently updated and contain a store of information both technical and practical.

ABC and XYZ of Bee Culture. A. L. Root Co., Medina, Ohio, 1944.

In spite of a certain pretentiousness, this is perhaps the most widely informative book on bees and beekeeping. It is alphabetically arranged (hence its unfortunate title), and constitutes an encyclopedia of beekeeping, with entries varying from a few lines to complete essays. Coverage is somewhat uneven, and sometimes gives the impression that not a great deal of actual updating went into some of the revised editions. An incredible number of copies have been sold over the course of half a century, some two hundred thousand, which is revealing as an index to its apparent authority as well as to the popularity of or interest in beekeeping itself. It must also set some kind of a record for a private press.

John Compton. *A Hive of Bees.* Doubleday & Co., Inc., Garden City, 1958.

This is an informative and often amusing account by a British amateur beekeeper, chronicling his beginnings as a hobbiest and his first experiences with bees. His interests sometimes become moderately technical, but a great deal of information about bees is presented in a more easily understood style and manner than will be found in more technical books. Even one with no interest in bees would probably enjoy this book, if only for the author's wryly British humor.

C. P. Dadant. *First Lessons in Bee Keeping. American Bee Journal,* Hamilton, Illinois, 1917. Revised 1962.

This book is often supplied to beekeepers who order a "complete beginner's outfit." It is very useful, but in spite of the recent revisions, seems awkwardly dated, even troublesomely out of date. As with many books which offer elementary instructions about a complex "practical science," it sometimes tends to tell you too much, or else too little, about crucial or confusing matters.

Arthur M. Dines and Stephen Dalton. *Honeybees from Close Up.* Thomas Y. Crowell Co., New York, 1968.

This book contains an absolutely marvelous collection of close-up photographs of bees, provided by Mr. Dalton. It also provides a quite readable text, but this is largely supplemental to the pictures. While it deserves reading, the text is not and does not intend to be a guide to cultivation of bees.

H. Malcolm Fraser. *Beekeeping in Antiquity.* University of London Press, Ltd., London, 1931. Second edition, 1951.

This is not, of course, a book of practical beekeeping. It traces the fascinating byways of ancient bee lore and superstition, as well as details the surprising state of science of ancient beekeepers. The book will provide a most pleasant winter read.

Roy A. Grout, ed. *The Hive and the Honeybee.* Dadant & Sons, Hamilton, Illinois, 1946.

This is a monumental book on beekeeping, the result of literally generations of revisions. One of its defects is that frequently it reveals its origins with anachronisms and consistently outdated pictures. The book contains twenty-five chapters, each written or compiled by an expert, covering phases of apiculture ranging from the practical and commerical to the technical and theoretical. It includes much information of use only to a professional or large-scale honey producer. Its usefulness to the hobbiest is unfortunately uneven, but as a reference work it is invaluable.

William Hamilton. *The Art of Bee Keeping.* The Herold Printing Works, York, England, 1951.

Like many beekeeping books by English practitioners, this will be found by a number of American readers to be frequently obscure or difficult to follow. If you can survive or tolerate its attitude of "I can tell from here you couldn't survive five minutes at a British Beekeeper's Convention," the book has its instructive moments. Mr. Hamilton's discussion of swarm control and prevention (where lucid) is particularly valuable.

F. N. Howes, D.Sc. *Plants and Beekeeping.* Faber and Faber, Ltd., London, 1945.

While chiefly an account of British flowers and circumstances affecting beekeeping in Britain, the book is nevertheless of considerable interest and value to American beekeepers. After preliminary essays on the various importance of plants to beekeepers, there is a section on major honey plants, describing in some detail the usefulness of most. Over half of the book is an alphabetically arranged listing of various plants visited by bees.

Murray, Hoyt. *The World of Bees.* Coward McCann, Inc., New York, 1965.

This account is directed to the layman, but is somewhat uneven in directing a beginner with bees exactly how to proceed. He includes an especially interesting chapter on bee venom and the allergies to bee stings, and anyone interested in bee stings and arthritis will be interested in his analysis of the question. His chapter on wintering bees is also particularly helpful.

Carl E. Killion. *Honey in the Comb.* Paris, Illinois, 1951.

If you are interested in producing the small sections of comb honey, special techniques and equipment are necessary. This book is very instructive, and ample photographs supplement explanations which are otherwise rather awkward.

Harry H. Laidlow, Jr., and J. C. Eckert. *Queen Rearing.* Dadant & Sons, Hamilton, Illinois, 1950.

There are many books and Department of Agriculture technical bulletins on the subject of queen rearing. This is an excellent guide by itself—readable, informative, and practical. It suggests programs of queen management and stock improvement which many advanced hobby beekeepers will be attracted to.

Allen Latham. *Allen Latham's Bee Book.* Hale Publishing Company, Hopeville, Georgia, 1949.

There is a great deal of useful information to be found in this book, which reflects the extensive career of its author as a professional beekeeper and enthusiastic student and close observer of bees. What he says is so, one is disinclined to doubt. The book is sometimes excessively anecdotal—perhaps an inevitable defect of what are sometimes the major virtues of the book—and it is sometimes querulous. His explanations are not always clear, particularly when he is trying to explain how to make something—his own preferred design for a beehive, for example. His discussion of the activities of the bees in the hives is particularly good. The book as a whole, however, is not a sufficient guide for an amateur beginning with his first hive of bees.

John A. Lawson. *Honeycraft*. Chapman & Hall, Ltd., 1948.

This book is characterized by the rather crotchety tone of the British amateur. It is sometimes informative, frequently opaque and difficult to follow. Its British terminology and reference to British beekeeping materials and styles of equipment further contribute to its tendency to confusion. At least for Americans, it is not to be ranked among the most useful works of reference, but if you want to see how the British amateurs do it, or at least write about it, this book should satiate your curiosity.

R. O. B. Manley. *Honey Farming*. Faber & Faber, Ltd., London, 1944.

Reading this book should be reserved as a treat to be indulged in after sampling two or three other British beekeeping books. It is a perfect foil to Mr. Lawson's book. Manley humbugs much of the doctrine of British amateurs, particularly the multitudes of different British hive designs which British authorities discuss so earnestly. He writes with grace, clarity, and modesty. His chapter on breeding bees is especially interesting, and is recommended to anyone who would like to build up his own stock rather than purchase queens from commercial stock.

Margaret Warner Morly, *The Honey Makers*. A. C. McClurg & Co., Chicago, 1915.

Although this book is utterly out of date and unserviceable as a guide to practical beekeeping, it is delightfully pregnant with quaint and curious lore and other matters associated with bees and honey. It has a very interesting chapter on mead, replete with numerous literary allusions, recipes, etc. The book makes very pleasant company.

Frank C. Pellett. *American Honey Plants.* American Bee Journal, Hamilton, Illinois, 1923.

Although somewhat dated, this book remains the most thorough study of the subject, and is exhaustively informative. It is encyclopedic, organized alphabetically by plant name, state, and topic. Most dated of the materials are the statements on honey crops relevant to the various states. However, it will tell you the nectar- and pollen-yielding characteristics of nearly any honey plant you might be curious about, as well as the quality and characteristics of most honeys produced.